# ETHNIC
# MINORITIES

# ETHNIC MINORITIES

## Social Psychological Perspectives

JAN PIETER VAN OUDENHOVEN
*Groningen University*

TINEKE M. WILLEMSEN
*University of Amsterdam*

Editors

SWETS & ZEITLINGER B.V. AMSTERDAM / LISSE     PUBLISHERS

SWETS NORTH AMERICA INC. BERWYN, PA     1989

*Library of Congress Cataloging-in-Publication Data*
Ethnic minorities /

   Bibliography: p.
   Includes indexes.
   1. Ethnic groups. 2. Minorities. 3. Ethnic relations. I. Oudenhoven,
Jan Pieter van, 1944-        II Willemsen, Tineke M., 1946-
   GN495.4.E844   1989              305.8                    89-11275
   ISBN 90-265-0988-X


*CIP-gegevens Koninklijke Bibliotheek, Den Haag*

Ethnic

Ethnic Minorities : social psychological perspectives /
Jan Pieter van Oudenhoven, Tineke M. Willemsen (eds.).-
Amsterdam(etc.): Swets & Zeitlinger; Berwyn: Swets North America
Met lit. opg., reg.
ISBN 90-265-0988-X geb.
SISO 328.7 UDC 316.6:3-054.62 NUGI 714
Trefw.: ethnische minderheden ; sociale psychologie


Cover design and layout: Rob Molthoff
Cover printed in the Netherlands by Casparie, IJsselstein
Printed in the Netherlands by Offsetdrukkerij Kanters B.V., Alblasserdam


ISBN 90 265 0988 X
NUGI 714

# CONTENTS

# FOREWORD

In this book we present an overview of the contribution social psychology may offer to the issue of interethnic relations. This book comprises three different approaches: social psychological theory; analysis and description of interethnic relationships in real life settings; and the application of theories to reduce discrimination.

The theoretical chapters of this book are primarily based on social categorization theory, but new developments such as social attribution theory and theories of emotions toward outgroups are also represented. As such, the present volume offers some new theoretical perspectives.

In the second part of the book the consequenses of prejudice or other forms of intergroup feelings in concrete settings are depicted. These chapters not merely present reports of empirical findings, but include careful analyses of, for example, conversations, classroom discussions, or police procedures. They allow us to gain a detailed insight into how the processes described in the theoretical chapters work out in daily life.

A substantial part of this book is dedicated to studies of remedies for problems of racism and discrimination. We do not pretend that this book offers a solution to all these problems, as we must conclude that, unfortunately, nor the state of knowledge nor the nature of current social psychology is such that large-scale solutions are available. However, these chapters do offer a number of practical suggestions for specific situations like the classroom and the working place.

The book is meant both for psychologists and sociologists involved in the issue of interethnic relations, and for students with similar interests. As it gives an overview of current theorizing on relations between minorities and majority, and of ways to attack real life problems, it seems especially suitable for a graduate course in social psychology.

We would like to thank Ms. M. Mitchell and Ms. H. Borkent who improved considerably the chapters written by non-native speakers of English. We are also very grateful to Fop Coolsma who did the tedious job of combining the separate references into one bibliography, and helped to compose the subject index.

Groningen, the Netherlands                    Jan Pieter van Oudenhoven
Amsterdam, the Netherlands                        Tineke M. Willemsen

March, 1989

# List of Contributors

WIM BERNASCO

Department of Clinical, Health, and Personality Psychology, Leyden University, The Netherlands.

ANTON J. DIJKER

Department of Social Psychology, University of Amsterdam, The Netherlands.

WILLEM DOISE

Faculté de Psychologie et des Sciences de l'Education, Université de Genève, Switzerland.

MILES HEWSTONE

Department of Psychology, University of Bristol, England.

WIEBE DE JONG

Department of Social Psychology, Faculty of Law, Erasmus University, Rotterdam, The Netherlands.

FABIO LORENZI-CIOLDI

Faculté de Psychologie et des Sciences de l'Education, Université de Genève, Switzerland.

JOANNE MARTIN

Graduate School of Business, Stanford University, Stanford, California, U.S.A.

ROEL W. MEERTENS

Department of Social Psychology, University of Amsterdam, The Netherlands.

THOMAS F. PETTIGREW

Department of Social Psychology, Unversity of Amsterdam, The Netherlands; and University of California, Santa Cruz, U.S.A.

SAWITRI SAHARSO

Centre for Race and Ethnic Studies, University of Amsterdam, The Netherlands.

HENRIETTE VAN DEN HEUVEL

Department of Social Psychology, University of Amsterdam, The Netherlands.

TEUN A. VAN DIJK

Department of General Literature, University of Amsterdam, The Netherlands.

AD VAN KNIPPENBERG

Department of Social Psychology, University of Nijmegen, The Netherlands.

JAN PIETER VAN OUDENHOVEN

Department of Social Psychology, Groningen University, The Netherlands.

ELS C. M. VAN SCHIE

Department of Social Psychology, University of Amsterdam, The Netherlands.

TINEKE M. WILLEMSEN

Department of Social Psychology, University of Amsterdam, The Netherlands.

# 1

# SOCIAL PSYCHOLOGICAL PERSPECTIVES ON ETHNIC MINORITIES: AN INTRODUCTION

Tineke M. Willemsen
University of Amsterdam
The Netherlands

Jan Pieter van Oudenhoven
Groningen University
The Netherlands

The 1980s seem to be the decade of ethnic unrest. Violence has increased in 'traditional' conflict areas, such as Israel, South Africa and Northern Ireland. Other regions, however, have also become reknowned for their problems relating to ethnic minorities: The Sikh from the Punjab and the Tamils from Sri Lanka are ethnic minority groups frequently mentioned in western news bulletins. All over the world there are conflicts between ethnic groups. Sometimes an ethnic minority dominates a majority, as in the well-known example of South Africa, where the political system is structured upon racial differences. More common are conflicts in which an ethnic minority is in one way or another discriminated against by an ethnic majority. We use the term 'ethnic minority' to define social groups that differ from the majority of the people in the country or society in which they live. Differences may refer to language, race or religion or a combination of these characteristics. However, conflicts between ethnic groups may have little to do with these differences; they may instead be related to, for instance, differences in power or a conflict of interests.

Relations between majority and minority groups are not necessarily problematic. In Switzerland, for instance, relations between the German-speaking majority and the French, Italian and Retroroman minorities are rather harmonious. Neither do Protestants living among a predominantly Roman Catholic majority in France experience major problems. Quite often, however, particularly when minorities do less well economically (or, conversely, are economically more successful than the majority), problems of prejudice, discrimination and racism arise.

The social psychology of interethnic relations has been developed primarily in the U.S. This book presents mainly European experiences and viewpoints. However, European intergroup situations and problems are not so different from those in the U.S., although the historical background and social context are different. Prejudice and discrimination against ethnic minorities in Europe are similar to what happens elsewhere. Before starting with a sketch of some relevant theories on the origins of prejudice, we first describe the main features of the Western European context.

## THE WESTERN EUROPEAN CONTEXT

Europe has always been very heterogeneous, ethnically and especially linguistically. In Europe dozens of official languages are spoken. Many ethnic minorities have existed for a long time, living among a majority, without major intergroup problems. Others like the Basques in Spain or the Roman Catholics in Ulster have known a problematic existence. In this book we will discuss primarily ethnic minority situations that have developed after World War II. Rapid changes in the ethnic situation in many countries have caused a renewal of scientific interest in the study of ethnic relations and ethnic conflicts.

In the post-war period there has been an influx of two major groups of people from other countries and other cultures into several Western European countries. The first group consists of people who emigrated from the former colonies to the 'mother country'. Some of them hoped to find better economic and educational opportunities in Western Europe: more jobs, and better schools for their children. Others emigrated because they had more or less identified with the colonisers, and no longer felt comfortable, or felt even unsafe, in their own decolonized homelands. These former colonial ties are clearly visible in the immigration streams into England, where people from Commonwealth countries like India and Jamaica came to live; in France and Belgium which attracted immigrants from African countries like Cameroon and Zaïre; and in the Netherlands, where many people from Indonesia, and later from Surinam, came to live when these countries became independent. A first characteristic these groups have in common is that, in general, they do not plan to return to their original countries; they plan to stay in Europe. A second common characteristic is that they generally were familiar upon arrival with the language of their new country, because it was the language taught in the colonial schools.

In both features they differ from the second post-war group of ethnic minorities who immigrated into Western Europe in the 1960s. During this period of economic growth, some industrialized countries started to recruit workers from the relatively poor Mediterranean countries. In Switzerland, Germany, the Netherlands, and Belgium many people from southern Italy, Turkey and Morocco arrived in this way. Even after the official recruitment programs were discontinued because of growing unemployment problems in the wealthy countries, the immigration flow did not stop. Immigrants continued to feel they would be better able to earn a living in these richer countries than in their home countries. In general, however, they did not see themselves as immigrants; they considered their stay a temporary one, and planned to return to their native countries when circumstances would permit. The children of these 'foreign workers', however, have grown up in Western European countries, speak the local language learned at school, and often do not intend to return to the country in which their parents were born. Thus, certain differences between the second generations of the two types of ethnic minority groups, the children of the immigrants from the former colonies and those of the workers from the Mediterranean countries, are gradually disappearing, with respect to such characteristics as mastery of the dominant language and willingness to stay in the countries to which their parents emigrated. However, there still are important differences between the two groups. On the whole, more members of the first group belong to the middle classes, a greater percentage of them hold jobs and they are more assimilated into the Western European culture.

The situation described above is clearly different from that in the United States. First, the native inhabitants of the U.S. nowadays constitute an ethnic minority. Second, the people who are considered to belong to the ethnic majority in studies in the U.S. have only their (white) skin color in common, not their national or linguistic origin: they are the descendants of European immigrants from such distinct countries as England, Germany, Poland, Ireland, the Netherlands, or France. Third, the largest minority group, the black Americans, did not originally immigrate voluntarily, but were forcefully transported from Africa to serve in slavery as a labor force. Another difference between black Americans and ethnic minorities in Europe is that black Americans have been living in the United States for almost as many generations as white Americans; the minority-majority situation has existed for centuries. More recent is the influx of Hispanics, especially Mexicans, whose position is comparable to those of the Mediterranean groups in Western Europe: their main purpose is to earn a better living than would be possible for them in their native country. Other new minorities in the United States come from countries which were involved in an armed conflict together with the United States, such as Korea or Vietnam. As a group, they have many of the characteristics we described of the immigrants from the former colonies into Europe. From a European viewpoint it is amazing that the social psychological literature is so predominantly concerned with the relationships between white and black Americans, and hardly at all with intergroup relations involving these more recently arrived ethnic minorities. The Hispanics form an extremely important ethnic group, not only for demographic but also for linguistic reasons: according to recent estimates 50 to 70 million Americans will have Spanish as their native language by the year 2000 (see e.g. Glazer, 1983).

The fact that the American ethnic situation differs from the European does not mean that the social problems ethnic minorities encounter in Europe are very different from those in the U.S. Ethnic stereotypes and racism have developed, and discrimination, both direct and indirect, does occur in Europe as well. For instance, in the Netherlands, the majority of students attending certain schools in the inner city of large cities are ethnic minorities, with the result that parents of native Dutch children have begun to remove their children to other, 'white' schools. This reminds one of the school desegregation problems in the U.S. in the '50s and '60s. Apparently, new forms of discrimination are created as the white majority sees the ethnic minorities grow. In the United Kingdom, for example, new immigration laws were introduced in 1971 that are racist in effect. Since 1971, legislation has been based on the concept of 'patriality'. Patrials are persons with a parent or grandparent of UK citizenship or who were born, naturalized or registered in the UK. Only patrials are allowed to settle. They come primarily from 'white' countries such as Canada, Australia and New Zealand. Societal developments such as these have led to a renewed interest of Western European social psychologists in the psychology of intergroup relations.

BASIC CONCEPTS

We first give a short definition of the basic concepts that will frequently be used in this book. *Prejudice* is a negative attitude toward a social group and toward individual members of that group. Unlike prejudice, a *stereotype*, which is a set of beliefs and expectancies about a social group, is not necessarily negative. Both prejudice and stereotypes are based upon a categorization process, a grouping of persons into categories on the basis of some common characteristic. *Discrimination* is a behavioral consequence from such a categorization; it consists of unequal treatment of (groups of) people belonging to a certain social category. Discrimination is generally defined as a negative behavior: denying persons certain rights merely on the basis of their belonging to a certain social category. *Racism* is a philosophy expressing the superiority of one race over another race. It may be expressed at an individual, an institutional, or a cultural level (Dovidio & Gaertner, 1986, p.3). Racism always implies prejudice and may encompass discrimination as well. The relations between these different concepts, and especially between cognitive structures such as prejudice and stereotypes on the one hand, and discrimination on the other hand, are generally not straightforward or simple. Most theoretical approaches assume a connection between prejudice and discrimination, thus between cognitions and behavior. Although virtually all authors state that by measuring prejudice one cannot predict directly the amount or the direction of discrimination, there is implicitly a firm belief that those who discriminate against certain groups are prejudiced or that those who are prejudiced will discriminate. However, the evidence for the relationship between situational cognitions, i.e. cognitions based on social norms and laws, and behavior is considerably stronger and more consistent than for the relationship between individual cognitions and intergroup behavior (Stephan, 1985, p. 632). We will return to this important aspect of the relationship between prejudice and behavior in the last chapter, where remedies are discussed, because it has important consequences for the ways in which discrimination and racism can be diminished. Suffice it to say here that the supposed relationship between individual prejudice and behavior has guided much work aimed at studying the origins of interethnic conflicts.

THEORIES OF PREJUDICE

Most psychologists assume that prejudice is the basis of racism and discrimination, and therefore theories that try to explain how intergroup conflicts and discrimination can occur are often theories of prejudice. The following brief sketch of the most relevant theories of the origins and functioning of prejudice is provided to enable the reader to place the theoretical viewpoints expressed in this book in a broader theoretical context. Special reference will be given to the contact hypothesis, social identity or social categorization theory, and social attribution theory. For more detailed accounts on theories of intergroup relations we refer to Katz (1976), Simpson and Yinger (1985), Stephan (1985) or Dovidio and Gaertner (1986).

In his influential book 'The nature of prejudice' Allport (1954, ch. 13) distinguished six theoretical approaches to the origins of prejudice. They are ordered on a continuum from very general influences on prejudice to direct causal influences on the act of discrimination. Two of them, the historical approach and the sociocultural approach, take the larger societal level (i.e. historical developments, the total social and cultural context) into account. Such macro-approaches are hardly used anymore in social psychology. Current theorizing moves away from societal factors toward individual cognitive factors. Nonetheless, some sociocultural factors mentioned by Allport, such as contacts between groups and social mobility of groups, still form part of many modern social psychological theories of prejudice.

A third theoretical approach in Allport's overview is the situational approach. This approach stresses immediate situational forces which influence intergroup contact and interethnic attitudes. The best known theory using this approach is the *contact hypothesis*, stating that, if some conditions are met, contact between different social groups will lead to a reduction of prejudice and discrimination, and to the development of friendly attitudes. Actually, it is not a theory of the origins of prejudice; it is a theory of prejudice reduction. Recent versions of this approach deal with the conditions in which intergroup contact leads to a decrease of prejudice. In the present book the chapter by Pettigrew and Martin and the chapter by Van Oudenhoven deal with aspects of the contact hypothesis, the conditions that can lead to succesful (i.e. prejudice-diminishing) contacts between ethnic groups.

The other three theoretical approaches discerned by Allport (the stimulus object, the psychodynamic, and the phenomenological) emphasize, to various degrees, individual psychological processes. The stimulus object approach, frequently and more clearly called earned reputation theory, assumes that prejudice is based on actual differences between groups which lead to dislike, discrimination and hostility. This model is now obsolete, in so far as it is a blame-the-victim approach.

The psychodynamic model emphasizes personality dispositions. This approach, which became influential with Adorno's concept of the authoritarian personality syndrome (Adorno *et al.* 1950), postulates that prejudiced persons are more close-minded, intolerant of ambiguity, anxious, and dogmatic. Very popular was the scapegoat theory based on the frustration-aggression hypothesis (Dollard *et al.*, 1939): frustration, for instance caused by unemployment, leads to aggression, and if that aggression cannot be acted out toward the persons responsible for the frustration it will be displaced toward any other victim. Although the intensity and content of prejudices can vary from person to person, most researchers agree that there is so much resemblance in racial stereotypes and prejudice within social groups that it cannot be satisfactorily explained at an individual level only. As is the case with other psychodynamically oriented theories in social psychology, this theory has become less important. It was more useful as an explanation for sudden outbursts of discrimination than for long-standing prejudice against specific groups. However, the scapegoat theory is still frequently used by laymen, e.g. journalists, when explaining incidents of ethnic unrest.

The phenomenological approach, as described by Allport, deals with an individual's way of thinking, stating that someone's conduct proceeds from the way he or she interprets the situation. In this approach stereotypes play a prominent role. This approach is clearly recognizable in most of the modern social perceptual theories of intergroup behavior, which were not yet developed when Allport made his classification. The most influential cognitive theory is *social categorization theory or social identity theory* (Tajfel, 1968, 1970; Tajfel & Turner, 1979, 1985). According to social categorization theory, an individual's identity depends to a large extent on social group memberships. People seek positive social identity, i.e. they tend to define themselves in terms of positive group memberships. Social comparisons between groups are basic to the evaluation of social identity. In order to achieve positive social identity intergroup comparisons are focused on the establishment of positive distinctiveness. Another assumption of social categorization theory is the perceptual accentuation of differences between people belonging to different social groups or categories, and at the same time the perceptual accentuation of similarities between people belonging to the same social group or category (Tajfel, 1981). The comparison sometimes refers to personal attributes such as 'intelligent', 'lazy', or 'dishonest' which have become subjectively associated with some social group through personal or cultural experiences. If such a classification is in terms of racial, ethnic, national or some other social criterion then it is likely that the categorization process is responsible at least in part for the biases found in the judgements of individuals belonging to various human groups (Tajfel, 1981). Social categorization theory is a cognitive theory and conseqently emphasizes intra-individual cognitive processing. However, its focus on group categorization and the cultural context within which this takes place makes it an attractively broad theory. Tajfel has always stressed the importance of the social context, and also the social functions of perceptual schemata such as stereotypes. According to Turner (1981) the elements that affect social identity are not only intra-individual and interpersonal factors but also intergroup relations of status, power, and material interdependence.

A different type of cognitive explanation is the *social attribution theory*. An important explaining factor in this theory is the fundamental attribution error, the phenomenon that one tends to explain behavior by internal factors rather than by situational factors. Thus, the conditions of ethnic minorities, such as poverty, are caused by supposed character traits of the ethnic minority, for instance lack of intelligence or laziness, rather than by circumstances beyond its control, such as discrimination by the majority. In a sense, social attribution theory can be considered a modern, cognitive version of earned reputation theory, in which perceived instead of actual differences are considered to be the basis of prejudice. However, attribution theory goes much further in explaining the mechanisms through which prejudice is formed and maintained, by specifying under which conditions it is possible to make situational attributions (and thus not blame the minority member), and to generalize these situational attributions to the whole category, i.e. the whole minority group. Hewstone (chapter 2) represents the social attribution approach in the present volume.

## NARROWING FOCUS OF SOCIAL PSYCHOLOGICAL EXPLANATIONS

Social psychology is often defined as the study of individuals in their social context. Not surprisingly, most social psychological theories of prejudice use the individual level of explanation. However, this is somewhat contradictory, given the enormous societal problems that cause psychologists' concern with prejudice. This concern pertains not so much to the fact that some individuals are biased against other individuals with another skin color, language or religion, but to the fact that in certain societies prejudices are shared by the majority of the dominant group; and this may lead to widely accepted forms of discrimination, sometimes even institutionalized or legalized, against the same ethnic groups. This is what makes the topic so important and urgent. Both Allport (1954) and Katz (1976) pay much attention to this larger societal level of discrimination. For instance, Katz's book includes a chapter that gives an overview of legal remedies for racial discrimination (Hirschhorn, 1976). Harding *et al.* (1969), in their chapter on prejudice and ethnic relations in the second edition of the Handbook of Social Psychology, explicitly choose the social problem approach. In more recent overviews of research on prejudice the societal level seems to fall into the background. Stephan (1985) for instance, in his chapter on intergroup relations in the third edition of the Handbook of Social Psychology, focuses primarily on the cognitive information processing approach. This difference in approach between the second and the third edition of the Handbook seems indicative of the narrowing interest of social psychology in the problems of prejudice.

We regret the tendency to explain prejudice and discrimination on an individual level for two reasons. First, although every social psychologist is aware of the fact that both individual cognitive processes and cultural and societal factors are relevant for the study of racism and prejudice, these latter factors are often not taken into account. This implies that almost all social psychologists are working on theories which explain only half of the phenomenon of interest.

Of course, it is quite legitimate to study only part of a phenomenon; it is even necessary for complex social situations. But since virtually all studies focus on the same half of the phenomenon, the other half gets neglected. The recent social psychological literature on intergroup relations could easily give the impression that interethnic conflict is primarily a matter of cognitive processes. This is a theoretical loss. Second, not less important, the emphasis on cognitive processes makes it much more difficult to develop remedies of prejudice. It is easier to change situations than to change the way in which these situations are perceived. We will return to this latter problem in the concluding chapter, in which we discuss in more detail the possible strategies for diminishing prejudice that follow from the theoretical approaches.

ABOUT THIS BOOK

The present volume consists of three parts. In part I (Chapter 2-5) each chapter exposes a theoretical point of view. As the book reflects current theoretical work, it will be no surprise that mainly cognitive theories are represented: social attribution, social categorization and social identity theory. However, emotional aspects of stereotypes are discussed as well. Together, these chapters are illustrations of recent European theorizing about intergroup perception and behavior, with an emphasis on cognitive approaches, and point to areas deserving of further investigation.

Part II (Chapter 6-9) contains more descriptive chapters, in which relations between ethnic groups are studied in a particular setting. These intergroup relations are described on different levels, i.e. the general level of conversations about minority groups and the very concrete level of living next door to each other; and in formal and informal settings, such as the police station or a school. The situations depicted in these chapters form a cross-section of the situations in which interethnic relations actually occur.

Part III (Chapter 10-12) deals with strategies to reduce prejudice and discrimination. In these studies theoretical viewpoints and knowledge are applied to improve actual interracial situations. In a way, the chapters in Part III form an integration of the first two parts of the book.

In the concluding chapter, we review the foregoing chapters and present some connections between, on the one hand, theories of prejudice and discrimination and, on the other, strategies to reduce these phenomena. Finally, we discuss the question of assimilation versus pluralism, and present our own view.

# Part I

# Theoretical Approaches

# 2

# INTERGROUP ATTRIBUTION: SOME IMPLICATIONS FOR THE STUDY OF ETHNIC PREJUDICE*

Miles Hewstone
University of Bristol
United Kingdom

Intergroup attribution refers to how members of different social groups *explain* the behavior, outcomes of behavior, and the social conditions that characterise members of their own (the ingroup) and other (the outgroup) social groups. In other words, we are not interested in the explanation of the behavior of individuals as such, but in the explanation of the behavior of individuals who act as members or representatives of social groups (or are perceived in those terms). This chapter highlights the link between *intergroup attribution* and *ethnic prejudice* by looking at the role of attributions in the development, maintenance and reduction of intergroup conflict.

The first part of this chapter concerns the maintenance of intergroup conflict by explanations for the behavior of ingroup and outgroup members. The possible motivational bases of such attributions are then examined separately for members of majority/dominant and minority/dominated groups, by considering the group functions fulfilled by attributions. Cognitive bases of intergroup attribution are then examined, with a focus on the role of expectancies and schema-based attribution. The second part of the chapter explores the reduction of intergroup conflict, by investigating attributions for behavior that disconfirms expectancies. This leads to an analysis of models of schema change. The paper concludes by setting out a model of the role of attributions in intergroup conflict.

## THE DEVELOPMENT AND MAINTENANCE OF INTERGROUP CONFLICT

Rather than attempt to explain when, where and how intergroup conflict begins, the attributional approach seems most valuable in underlining how conflict is supported, even exacerbated. Konecni's (1979) analysis of the role of aversive events in the development of intergroup conflict stated that such events represent information that affects thoughts and attitudes concerning members of the outgroup, especially when an ingroup member perceives an aversive event to be due to the actions of another group. He reported that there are literally countless examples of aggressive behavior aimed directly at members of outgroups perceived to be responsible for the occurrence of aversive events (see Gurr, 1968). In a similar vein, Billig's (1976) account of the frustration-aggression hypothesis applied to intergroup relations argued that the crucial intervening variable that links the cause (e.g., frustration due to social deprivation) with the effect (e.g., aggression in the form of a riot) is the social interpretation of the deprivation. As Billig put it, the *theories of social causation* shared by the participants in such collective behavior are very important.

Starting from a more traditional attribution theory perspective, Cooper and Fazio (1979) extended Jones and Davis' (1965) Correspondent Inference Theory to intergroup conflict. Jones and Davis proposed a strong tendency to make more extreme dispositional attributions when the behavior of an actor directly affects the attributor in a positive or negative sense (hedonic relevance; see Chaiken & Cooper, 1973) and when the behavior is seen to be intended for the attributor (personalism). Cooper and Fazio proposed that both tendencies were heightened in intergroup conflict, especially for negative actions by the outgroup, since these actions may be aimed at frustrating the aspirations of the ingroup. These simplified inferences about outgroup hostility may also enhance feelings of control, because counter-action against the outgroup appears the clear solution. Cooper and Fazio offered the interesting suggestion that personalism may only rarely influence our behavior as individuals, but that it may have a stronger effect on intergroup encounters. They introduced the term 'vicarious personalism' for one group's perception that the other group's actions were aimed specifically *at them*. This perception results in a more negative evaluation of the outgroup than would be implied by a dispassionate inference process; in short, "A simplistic correspondent inference about the evil nature of the outgroup is made" (p. 152). If such attributional bias and distortion of evidence is characteristic of intergroup perception, then there should be clear differences in the explanation of ingroup and outgroup members' behavior. I now briefly consider some relevant evidence.

*Explanations for the behavior of ingroup and outgroup members*

The studies reviewed in this section show that the causes of the same behavior can be perceived in very different ways depending on who performs the behavior (Hewstone & Jaspars, 1982a, 1984). The first empirical study to explore intergroup attributions was by Taylor and Jaggi (1974), and was carried out in southern India against the background of conflict between Hindu and Muslim religious groups. The basic hypothesis of the study was that observers (Hindu adults) would make internal attributions for other Hindus (i.e., ingroup members) performing socially desirable acts, and external attributions for undesirable acts. The reverse was predicted for attributions to Muslim outgroup members. The predictions were clearly borne out by the data.

Because of the importance of this study, a conceptual replication of it was carried out some years later, in southeast Asia (Hewstone & Ward, 1985). A first study used Malay and Chinese students in Malaysia. Malays behaved as expected, by making internal attributions for the positive behaviors of their own group members, but for negative behaviors by the Chinese. This constituted clear evidence of ethnocentric attribution, with the effect for *ingroup favoritism* actually far stronger than that for *outgroup denigration*. The Chinese, however, favored the Malay actors at the expense of their own group. This finding runs contrary to Taylor and Jaggi's (1974) predictions, but is consistent with previous findings that members of some minorities devalue their own group (e.g., Doise & Sinclair, 1973; Lambert *et al.*, 1960). A second study was then carried out in neighboring Singapore. The Malays retained the tendency to make internal attributions for positive behavior by their own group, but they did not make significantly different attributions for positive and negative behavior by the Chinese. The Chinese did not significantly favor either group.

There are many differences between Malaysia and Singapore (see Ward & Hewstone, 1985), but these different sets of results can be seen in terms of the different levels of inter-ethnic group conflict in the two countries, with more conflict in the politically rather tense and potentially assimilationist culture of Malaysia, than in relatively tolerant, multicultural Singapore. These attributional data were backed up by ethnic stereotypes. In Malaysia, Malays saw themselves in positive terms, but viewed the Chinese in predominantly negative terms (as did the Chinese themselves). The Singaporean data revealed a striking overall decrease in stereotyping. The responses of the Chinese in Malaysia may also be due to the rather distinctive role many of them enact in Malaysian society. They constitute a prototypical "middle man minority" (Blalock, 1967), occupying a marginal role between producer and consumer as a response to hostile reactions from the surrounding community (Bonacich, 1973). Bonacich argued that despite their undeniable economic successes, which should engender pride in group membership, "Discrimination and hostility against minorities usually has the effect of hurting group solidarity and pride, driving a group to the bottom rather than the middle of the social structure" (p. 584). Our results are therefore consistent with a large sociological literature.

A striking example of ethnic bias in attribution was provided by Duncan's (1976) study. Duncan examined the perception and explanation of intra- and inter-racial violence by asking subjects (white American College students) to look at a videotaped interaction of an increasingly violent argument in which, finally, one participant pushed the other. Duncan varied the race (black/white) of both the potential 'protagonist' and 'victim' of the push shown on the videotape. Viewers were first asked to *describe* what they saw, using categories such as 'playing around' and 'violent behavior'. The results were clear-cut. When the videotape contained a black protagonist (and no matter what race the victim was), over 70% of the subjects chose 'violent behavior' as the appropriate category. But, when the roles were reversed, and the protagonist was white, only 13% of the subjects labelled the act in this manner. Duncan then went a stage further in his experiment. He asked his subjects to *explain* the observed behavior. Again the results showed a clear effect for race of protagonist. When the protagonist was black, subjects said that the violent behavior was due to personal characteristics of the harm-doer; when the protagonist was white, on the other hand, subjects 'explained away' the behavior in terms of the situation.

Although the studies by Duncan (1976), Taylor and Jaggi (1974) and Hewstone and Ward (1985) demonstrated the phenomenon of inter-group attribution they are all limited in one important respect. They centred on the simplistic, if classic, dichotomy between internal and external attributions. An improved and multi-dimensional approach to the structure of perceived causality has been put forward by Weiner (1979, 1983, 1986). Weiner specifies the underlying proper-ties of causes in terms of three dimensions. *Locus* refers to the famil-iar location of a cause, internal or external to the person; *stability* refers to the temporal nature of a cause, varying from stable to unstable; and *controllability* refers to the degree of volitional influ-ence that can be exerted over a cause, ranging from controllable to uncontrollable. The value of this approach can be seen by looking more closely at the four attributions typically given for success and failure in academic settings – ability, effort, luck and the task – and considering how they might be used to explain success and failure by members of the in- and outgroup (see Table 1). The first thing to note from Table 1 is that there are multiple possibilities for explain-ing *outgroup success* and *ingroup failure* in group-serving terms. The second thing to note is that, contrary to Taylor and Jaggi, attribu-tions for ingroup-failure and outgroup-success can both be *internal*, and yet group-serving. According to Taylor and Jaggi, both attribu-tions should be *external*. The key here is that while ability and effort may both be seen as internal causes of achievement, ability is further classified as stable and uncontrollable, while effort is unstable and controllable. This, presumably, gives quite a different meaning to the two 'internal' attributions. Using this approach one can extend Taylor and Jaggi's model of intergroup attribution and apply it to achievement situations (see Deaux & Emswiller, 1974; Greenberg & Rosenfield, 1979; Hewstone, Jaspars & Lalljee, 1982; Yarkin, Town & Wallston, 1982).

*Table 1.* Ingroup-serving and outgroup-derogating attributions in achievement contexts.

| Type of outcome | Type of actor | |
| --- | --- | --- |
| | Ingroup | Outgroup |
| Success | Ability (internal, stable, uncontrollable) | Effort (internal, unstable, controllable) |
| | | Luck (external, unstable, uncontrollable) |
| | | Task (external, stable, uncontrollable) |
| Failure | Luck (external, unstable, uncontrollable) | |
| | Task (external, stable, uncontrollable) | |
| | Effort (internal, unstable controllable) | |

Note. Classification of attributions based on Weiner (1979).

The study most relevant to ethnic prejudice was carried out by Greenberg and Rosenfield (1979). These authors questioned whether intergroup attributions were based simply on dislike for outgroup members (ethnocentrism), or whether they were always founded on cultural stereotypes. To examine this question with respect to inter-racial (black-white) attributions, these researchers took a task, extra-sensory perception, about which no race-based cultural assumptions existed. In addition, white subjects of varying degrees of ethnocentrism were used. The subjects watched four videotapes portraying success and failure for both black and white actors and attributed each performance to Weiner's four specified causes.

In the case of success, highly ethnocentric subjects tended to attribute the performance of blacks less to ability and more to luck than the performance of whites; low ethnocentric subjects attributed blacks' behavior more to ability and less to luck than whites' behavior. For the failure outcomes, highly ethnocentric subjects ascribed the blacks' performance more to lack of ability than that of whites; while low ethnocentric subjects attributed the performance of blacks less to lack of ability and more to luck. These results were interpreted as evidence of intergroup attribution biases based on ethnocentrism alone, although it may still have been the case that the highly-ethnocentric subjects (the only subjects who showed intergroup attributional bias) did believe blacks to be inferior on this task.

The most troubling results are those indicating that outgroup success is not taken at face value, but is rather 'explained away'. This finding has been treated in more detail by Pettigrew who proposed an "ultimate attribution error" which is "a systematic patterning of intergroup misattributions shaped in part by prejudice" (1979, p. 464). Pettigrew argued that positive acts by members of disliked outgroups could be attributed to any one or the combination of the following:
A. to the exceptional, even exaggerated, special case individual who is contrasted with his/her group; B. to luck or special advantage which is often seen as unfair; C. to high motivation and effort; and/or D. to manipulated situational context. (Pettigrew 1979:469.)

This type of attribution may have depressing implications for the impact of affirmative action programmes, aimed at improving educational and employment opportunities for ethnic minorities and women. In order to achieve this goal, minorities may be given preferential treatment in job placements and promotions. However, at least one study has shown that minority students admitted to graduate school were judged *less* qualified when they were described as being on an affirmative action programme, than when not (Garcia *et al.*, 1981). It would seem likely that any successes achieved by such students might also be explained away in the terms suggested by Pettigrew (see also Chapter 10 in this reader).

There is, then, clearly evidence of intergroup attributional biases. The next two sections explore both motivational and cognitive bases of intergroup attribution, as well as point to some of the affective consequences of attribution.

*Motivational bases and affective consequences of intergroup attribution*

Weiner (1982; see also 1985) wrote that attributions appear to be sufficient antecedents for the elicitation of a number of emotions, including anger, pride (self-esteem), and resignation. He also argued that the underlying dimensions of attributions were significant, and sometimes necessary, determinants of these affective reactions. To consider the emotional consequences of intergroup attributions, one has to generalize from Weiner's work; in this section I consider the consequences of attributions separately for majority, or dominant, and minority, or dominated, groups.

*Majority groups.* Following Weiner, pride and positive self-esteem are experienced as a consequence of attributing a positive outcome to the self, whereas negative self-esteem is experienced when a negative outcome is ascribed to oneself. Thus one can see what emotional consequences follow from attributing both ingroup success and outgroup failure to internal, stable causes.

According to social identity theory (e.g., Tajfel & Turner, 1979), individuals define themselves to a large extent in terms of their social group memberships and tend to seek a positive social identity (or self-definition in terms of group membership). I propose that intergroup attributional bias may function in this way for majority group members. Tajfel's (1969) cognitive analysis of prejudice made the point that an individual's system of causes must provide, as far as possible, a positive self-image. Such a possible motivational bias is essentially a group-based equivalent of individual self-serving biases in attribution. Thus while a tendency has been found for individuals to explain events in ways that serve their own needs by enhancing their personal identity (e.g., Weary, 1979; Zuckerman, 1979), group members also tend to explain events in ways that would enhance their social identity.

However, there are obvious gaps in the available research. There is a clear need for research which, for example, investigates whether certain patterns of intergroup attribution are associated with a more or less positive self-esteem/social identity (cf. Lemyre & Smith, 1985). In addition, given the availability of measures of social identity (Brown & Williams, 1984), one could examine whether those who identify more with their own group make more ethnocentric attributions.

*Minority groups.* Causal attributions can also be related to group-esteem for minority groups. However, a more interesting emotional consequence is, perhaps, *anger*, as experienced when a negative, self-related outcome or event is attributed to factors controllable by others (Weiner, 1982). For minority groups, an attributional approach can contribute to the neglected analysis of the experience of discrimination. Experimental work by Dion and colleagues (Dion & Earn, 1975; Dion, Earn & Yee, 1978) has shown exactly this attribution-emotion link. Lone minority group members (e.g., Jews) who attributed their failure in a group ticket-passing task to religious discrimination by gentiles, reported feeling more aggression (as well as, sadness, anxiety and egotism) than did subjects who had not made such attributions. Perhaps even more interesting, these subjects also rated themselves more favorably on positive traits underlying the Jewish stereotype (e.g., 'love of traditions', 'industrious', 'clever'). Thus failing, *if* one's failure is ascribed to discrimination by members of another group, can lead to a *strengthened* commitment to the ingroup (see also Turner *et al.*, 1984).

Another important type of attribution for minority group members is 'system blame' as in black people's explanations for what might appear to be personal failures (in the eyes of Whites). Hewstone and Jaspars (1982b) examined the explanations for institutionalized racial discrimination given by young blacks and whites in Britain. Black respondents generally attributed the phenomena of discrimination more to white members of the system, and less to personal characteristics of blacks, than did white respondents, a tendency that was polarized in group discussion. Simmons (1978) wrote that this system blame may be a very important reaction to life in a racist society, because it protects self-esteem. In a similar vein, Gurin, Gurin, Lao and Beattie (1969) reported on young blacks in the U.S. who felt that economic or discriminatory factors were more important than individual skill and personal qualities in explaining the problems they faced. At one level these young people might seem to have an external locus of control, a belief that rewards are not controlled by themselves (see Rotter, 1966). While it has usually been assumed that internal locus of control represents a positive orientation, Gurin et al. suggested that this is not so for people disadvantaged by minority status, for whom an internal locus might lead to self- (and ingroup-) derogation and blame. Gurin et al. showed that those young blacks who blamed the system, often aspired to jobs that were non-traditional for blacks, and they were more ready to engage in collective action.

Attributions would appear to play an important role in generating what Billig (1976) called an "ideology of discontent", whereby a powerful outgroup is perceived to be the cause of, e.g., racial discrimination (Caplan & Paige, 1968) and the dominated group develops a positive ingroup ideology and challenges the dominant group. Tajfel and Turner (1979) maintained that low status groups only challenge high status groups if they perceive the status relationship as illegitimate and unstable (see Turner & Brown, 1978). It is therefore interesting to note reported relationships between ingroup-serving attributions and measures of perceived illegitimacy and instability for minority groups (see Bond et al., 1985; Hewstone et al., 1983). However, it is still not clear whether attributions play the important causal role of determining such perceptions and, in turn, conflict, or whether they are reflections of developing intergroup conflict. Future work should therefore analyse the role of intergroup attributions in more detail, using correlational and causal modelling techniques.

A possible mediational role for attributions should also be considered in cases of 'asymmetrical' social perception, where members of minority groups evaluate themselves negatively and the dominant group positively (Schwarzwald & Yinon, 1977). Demonstrations of such attributional "self-hate" (Lewin, 1948) have been reported for women (see Deaux, 1984; but cf. Feldman-Summers & Kiesler, 1974) and for ethnic Chinese in Malaysia (Hewstone & Ward, 1985). In this relatively rare case, attributions would be dysfunctional for the ingroup, with emotions of helplessness and resignation engendered by the attribution of negative outcomes to internal, stable factors (Weiner, 1982).

*Cognitive bases of intergroup conflict*

*Knowledge based attributions.* Just as Weiner's work led to the consideration of motivational bases of intergroup attribution, so too it leads to a discussion of cognitive factors. Weiner *et al.* (1972) reported that attributions were made to stable and internal causes when there was a fit between expectancy and performance, while a performance discrepant with expectancies was attributed to unstable causes. At an intergroup level, Deaux (1976, 1984) has linked causal attributions (for the performance of males and females) to observers' initial expectations concerning performance. Thus a male's successful performance on a male-linked task was attributed more to ability than was a female's successful performance, by both male and female subjects (Deaux & Emswiller, 1974; but note that this study used only one rating, an ability-luck dimension).

Following Deaux's (1976) work, cognitive social psychologists (e.g., Hamilton, 1979) began to think of stereotypes as 'structural frameworks' or 'schemata' in terms of which information is processed. A schema has been defined as, "a cognitive structure that represents organized knowledge about a given concept or type of stimulus" (Fiske & Taylor, 1984, p. 140); this knowledge structure has implications for social information-processing, including *perception, memory* and *inference*. In explaining schema-consistent or expectancy-confirming behavior, perceivers may simply rely on dispositions implied by the stereotype, not even bothering to consider additional factors (Pyszczynski & Greenberg, 1981). The search for possible causes of an effect is terminated once an effect has been adequately explained.

Given the contemporary importance of the schema notion, it would be useful to have more data on intergroup attributions as schematic inferences. For example, do we take longer to explain a schema-inconsistent outcome such as success by a disliked outgroup member, or failure by an ingroup member? Some recent data reported by Hewstone, Benn and Wilson (1988) are suggestive of possible effects. Subjects were not asked explicitly to make a causal attribution, but to assess the likelihood, given identical information, that a West Indian or White person had been accurately identified as a burglar. Not only did subjects spend *less* time deciding in the former case, they also used the available information quite differently – seeming *prejudiced* against the West Indian but wanting to *exonerate* the White person.

*Salience effects.* Taylor and Fiske (1978) have integrated evidence from a number of empirical studies to propose that many perceivers seek a single, sufficient, and salient explanation for behavior, and that causal attributions are often shaped by highly salient stimuli. Taylor and Fiske's overall hypothesis is that attention determines what information is salient, and that perceptually salient information is overrepresented in subsequent causal explanations. For example, a single black person in a small group of otherwise white people should be salient to observers and also perceived as disproportionately causal in the group's performance (Taylor, Fiske, Etcoff & Ruderman, 1978). That 'solo' person's race is the basis for his or her distinctiveness, and that attribute will be highly available (Tversky & Kahneman, 1974) as an explanation for the solo black person's behavior. Furthermore, attention to these salient stimuli also leads perceivers to mistake members of a different group for each other (Malpass & Kravitz, 1969; Taylor *et al.*, 1978); if 'they' are all seen as similar, then 'their' behavior can be explained in the same manner across outgroup actors.

*An integration*

In view of the foregoing discussion of motivational and cognitive factors in intergroup attribution, an integration seems most plausible. Attribution theorists (e.g., Bradley, 1978; Miller & Ross, 1975) have discussed this issue in some detail, leading Tetlock and Levi (1982) to conclude that for the present it appears impossible to choose between the two viewpoints. Similarly, Turner (1981), in the area of intergroup relations, has argued that cognitive (social categorization) and motivational (social comparison) processes have complementary effects on intergroup differentiation. Perhaps the more important implication for studies of intergroup conflict is that such a variety of potential cognitive and motivational bases of intergroup attribution exist. This fact further underlines the importance of the phenomenon, particularly when outgroup behavior actually disconfirms expectations and thus has the potential to reduce conflict by changing intergroup perceptions. This topic is considered in the remainder of the present chapter.

BEHAVIOR THAT DISCONFIRMS EXPECTATIONS: IMPLICATIONS FOR
THE REDUCTION OF INTERGROUP CONFLICT

A particularly important issue for the reduction of intergroup conflict is how perceivers react to information about the outgroup that
disconfirms their negative expectations. For example, the 'contact
hypothesis' – the belief that positive association with persons from a
disliked outgroup will lead to the growth of liking and respect for
that group (Cook, 1978) – appears to be based, in part at least, on the
value of disconfirming negative expectancies about the outgroup.
An attributional analysis of contact research points clearly to some
of its pitfalls: researchers must ensure that counter-stereotypic
behavior cannot be 'explained away' in terms of situational
demands or individual exceptions to the rule (Hewstone & Brown,
1986; Pettigrew, 1979; Williams, 1964). How then is unexpected
behavior explained, and what implications does this have for the
reduction of intergroup conflict?

*Attributions for behavior that disconfirms expectancies*

What type of explanation is given for unexpected behavior? Kulik
(1983) has presented compelling evidence for the role played by
causal attribution processes in belief perseverance. He reported that
the perceived causal importance of situational factors was influenced by the degree to which an observed behavior was consistent
with prior beliefs about the actor. Behavior that was consistent with
prior conceptions was attributed to dispositional characteristics of
the actor. Furthermore, situational factors that would in other cases
be seen as compelling explanations, were ignored in favor of dispositional factors as causes of expected behavior.

Inconsistent behavior, in contrast, was apt to be situationally
attributed. Even settings normally considered to inhibit the
observed behavior were judged instead as causal. This pattern of
attribution provides a cognitive basis for the Taylor and Jaggi effect.
In addition, this confirmatory attributional tendency limits the
potential influence of belief-disconfirming or schema-inconsistent
information. Such a simple process would act to preserve and protect stereotypes about the outgroup. Because stereotypes refer to
perceivers' assumptions about the dispositional attributes of
ingroup and outgroup members, any behavior violating the stereo-

type could be avoided on the basis that it reflected situational influences and thus did not derive from the personal characteristics of the actor (see Hamilton, 1979). As Merton argued, "the systematic condemnation of the out-grouper continues largely irrespective of what he does" (1957 edn., p. 428).

## Models of schema change

It is generally agreed that one of the main ways that schemata can change is through exposure to incongruent information (i.e., information that is improbable, given the schema). Weber and Crocker (1983) have compared three models of how beliefs such as stereotypes change in response to disconfirming information. The 'bookkeeping' model (Rothbart, 1981) views stereotype change as a gradual process in which each new instance of stereotype-discrepant information modifies the existing stereotype. Any single piece of disconfirming evidence elicits only a minor change; substantial changes occur incrementally with the accumulation of evidence that disconfirms the stereotype. The 'conversion' model (Rothbart, 1981) is more dramatic, allowing for a single, salient incongruent instance to bring about schema change. According to this second model, schema change is all-or-none and is not brought about by minor disconfirmations. The third model, 'subtyping', argues that when all the disconfirming information is concentrated within a few individuals, those individuals will be subtyped (seen as a separate subcategory). This model predicts more change when incongruent information is dispersed across individuals than when it is concentrated in a few (cf. Gurwitz & Dodge, 1977).

Weber and Crocker's (1983) four studies provided partial support for two of the models. The bookkeeping model may describe how stereotypes change when incongruent information is dispersed across multiple outgroup members (because in this condition individuals could not easily be subtyped, and the stereotype changed with each new piece of incongruent information). The subtyping model may be the best description of stereotype change when incongruent information is concentrated in a few outgroup members (the small number of individuals in whom information is concentrated are easily subtyped). Finally, although unsupported in this research, the conversion model might apply when a perceiver is unsure of a stereotype, and hence easily swayed by available information.

An attributional analysis of this research is quite informative. It argues, first, that both the bookkeeping and especially the conversion models run the risk that discrepant information will be explained away. Attributing discrepant information to, for example, some special personal characteristic of the actor implies that the experience of intimate contact (e.g., Rose, 1981) will generalize across situations, but not across other outgroup members. As Hewstone and Brown (1986) have argued, this failure to generalize attitude change is one of the central weaknesses of the traditional contact hypothesis. It is for this reason that we emphasized the interpersonal-intergroup continuum in our model of intergroup contact. The intergroup level of contact refers to behavior between individuals that is determined by their identity as group representatives, rather than by their individual characteristics and personal relationships. If positive interaction with a member of the outgroup takes place on an intergroup basis, then there is a real chance that any positive change in attitude will be seen to apply not just to that individual, but to others in the same group. In a similar vein, Rothbart and John (1985) proposed contact with a 'prototypical' outgroup member. Wilder (1984) has provided experimental evidence that a pleasant encounter with an outgroup member led to a significant improvement in the evaluation of the outgroup as a whole *only* when the target outgroup member was *typical*. Weber and Crocker (1983) also found that people's stereotypes about occupational groups changed most when they were presented with counter-stereotypical information about *representative* members of those groups. The same information, when associated with atypical members of the category in question, had much less effect in modifying attitudes.

One of the clearest lessons to emerge from the study of intergroup contact is that the experience of positive contact rarely generalizes to other outgroup members, because of the cognitive processes (including causal attribution) that classify the single outgroup member as a 'special case' and thus prevent belief change (see Allport's, 1954, "re-fencing device"; Williams's, 1964, "exemption mechanism"). Crocker *et al.* (1983) suggested that, in the absence of a dispositional attribution, people may conclude that incongruent behavior does not provide good evidence of the actor's typical behavior. In situations of intergroup contact, what is so often missing is information that leads the participants to view their partners as typical outgroup members, and thus to prevent them from explaining away positive behavior of the outgroup member (Hewstone & Brown, 1986).

The implications of this view are summarized in Fig. 1 in the form of a schematic model. The model is based on, but extends, previous work (Cooper & Fazio, 1979; Deaux, 1976; Hewstone & Brown, 1986; Stephan & Rosenfield, 1982; Taylor & Jaggi, 1974), and draws together the evidence I have reviewed concerning how attributions influence the development, maintenance and attempts at the reduction of conflict. The model assumes that there exist negative expectancies for outgroup behavior. These negative expectancies can be confirmed by one of two routes. First, outgroup behavior may be *perceived* to confirm expectancies and is thus attributed internally (to stable causes), maintaining conflict. Second, negative expectancies may lead to modifications of the behavior of ingroup members; outgroup behavior that is affected by the ingroup's expectancies; and outgroup behavior that *actually confirms* ingroup expectancies (a self-fulfilling prophecy). This behavior is again attributed internally, maintaining conflict.

The model also allows for outgroup behavior that does, or is perceived to, *disconfirm* expectancies. Unexpected behavior can be explained in various ways. First, it can be attributed to external factors, which again maintains conflict. Alternatively, the unexpected behavior can be attributed internally, with different consequences, depending on how the individual outgroup member is perceived. If perceived as *a*typical, the behavior can be 'explained away', by treating the individual as a 'special case'. This attribution means that the behavior has little impact on the outgroup stereotype. If, and only if, the individual is seen as a typical outgroup member, is there a real chance of generalized change of outgroup attitudes.

Such a model helps to explain why intergroup conflict often persists despite information that disconfirms negative expectancies about the outgroup. As the model shows, in two out of three cases where outgroup behavior is perceived to disconfirm expectancies, conflict-maintaining attributions can be given.

*Figure 1.* Attribution of outgroup behavior and the continuation or reduction of intergroup conflict (*after Hewstone, 1988*).

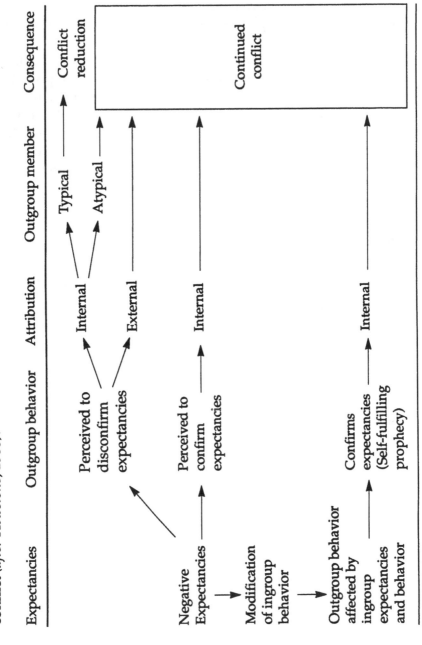

CONCLUSION

This chapter has reviewed some of the implications of an attributional approach to the phenomena of ethnic prejudice. There is clearly evidence of differential explanations for identical behavior by
ingroup and outgroup members, although this finding is not universal. Where found, however, it would seem to make a potent contribution to intergroup conflict. Intergroup attributions have viable
motivational and cognitive bases, as well as potential affective consequences for both majority and minority groups. Finally, the well-
documented attributional bias against behavior that disconfirms
negative expectations underlines the importance of intergroup attributions for the reduction of intergroup conflict.

*Note*

*This chapter is based on 'Attributional bases of intergroup conflict', which
appeared in W. Stroebe, A. W. Kruglanski, D. Bar-Tal and M. Hewstone (Eds.),
*The social psychology of intergroup conflict*. New York: Springer-Verlag (1988).

# 3
# PATTERNS OF DIFFERENTIATION WITHIN AND BETWEEN GROUPS

Willem Doise
Fabio Lorenzi-Cioldi
Geneva University, Switzerland

Most theories in the field of intergroup relations assume that within group solidarity increases during between group conflict or, more generally, that between group differentiation is inversely related to within group differentiation. In this chapter, we argue that both kinds of differentiation are bound together in more complex ways and that the possibility of a positive link should also be considered. Assumptions about such a link seem necessary in order to interpret the results of research carried out in the framework of different traditional theories of realistic intergroup conflict (Sherif, 1966), category differentiation (Tajfel, 1981) and social identity (Tajfel & Turner, 1986). More recent findings will be presented which illustrate specific processes generating social differentiation both among ingroup members and between members of different groups.

## LIMITATIONS OF THE SUMNER HYPOTHESIS:
## RED DEVILS, BLACK SHEEP AND FLEMINGS OF BRUSSELS

"The relation of comradeship and peace in the we-group and that of hostility and war toward other groups are correlative to each other."

This statement, quoted from Sumner (1906: 22), seems to have received much empirical support in intergroup research, which has often shown that an increase in intergroup competition and conflict goes together with a tightening of within group links (see Doise, 1978). Sherif's (1966) experiments, which took place in summer camps, are especially valid in illustrating such intergroup dynamics. They show that, when two groups are set to realize incompatible ends, such that one group cannot attain its goal unless the other fails to do so, an unfavorable perception develops between the groups, and the members of the one group can only think of and realize hostile contacts with the members of the other group. At the same time, within group solidarity increases and members adhere more strictly to ingroup norms, particularly to those bearing on relationships with outgroup members.

However, research also shows that competition between groups does not always make within group solidarity stronger; it can also enhance within group conflict, at least temporarily, as is reported in the following observation by Sherif and Sherif (1969: 64): "The victorious Bull Dogs were elated, happy, self-content, and full of pride. The losing Red Devils were dejected. Chiefly because their leader became vindictive, blaming defeat on low-status members of his own group, their loss was conducive to signs of disorganization. Low-status Red Devils resented the accusations, and there was conflict within the group until later the Red Devils faced broadside attacks from the Bull Dogs."

Intergroup competition can therefore generate strong intragroup conflict, which is also illustrated in Diab's (1970) replication of Sherif's experiments, in Lebanon. His competing groups were composed of both Christians and Muslims. His results showed many similarities to those of Sherif, but there were also differences: Only the winning group evaluated its performance as better than that of the other group and, at the end of the intergroup competition, four members of the losing group wanted to quit the camp, having developed strong aggression not only towards the members of the victorious outgroup but also towards one member of their own group. It is tempting to invoke the cultural context to explain these differences, but this would mean forgetting that Sherif also reported within group conflict during the competition between the Red Devils and the Bull Dogs. Moreover, if the Lebanese context did have some influence in Diab's experiment, we still need to define the mechanism which might have been operating. In this connection, we should recall the existence of crossed memberships, in so far as the two competitive groups were each made up of Christians and Muslims. This could explain why, at a given moment, sociometric choices were less influenced by experimentally produced group boundaries than in Sherif's experiment. It has therefore been claimed that, for the losing group, the difference in religious affiliation became salient: the four boys who wanted to quit the camp were all of the same faith, the definite failure in the tournament having eliminated all meaning from their group membership. It is apparantly a frequent observation among ethnologists (Beals, 1962; Chance, 1962) that groups which do not succeed in achieving a goal fragment into subgroups along prior demarcation lines.

An important question remains, however: do such within group cleavages appear only in losing groups or in groups which cannot achieve their basic goals? Recent research by Marques (1986) on 'the black sheep effect' describes a more general mechanism of within group differentiation in situations of intergroup encounters. Basically, the black sheep effect is also considered to be a consequence of group members trying to achieve superiority for their own group as compared to relevant outgroups. However, the hypothesis of Marques states that a way of achieving this superiority consists of downgrading marginal or deviant ingroup members who do not display the positive characteristics of the more prototypical ingroup members. Only these members are considered to be relevant for establishing a positive identity as compared to the outgroup.

Several experimental results reported by Marques (1986) illustrate the black sheep effect. In one experiment, Belgian students were asked to describe "likeable Belgian students", "unlikeable Belgian students", "likeable North African students" and "unlikeable North African students". As hypothesized, likeable Belgian students were described in a more positive way than likeable North Africans, while unlikeable Belgians were described in a more negative way than unlikeable North Africans. Another experiment showed that these effects occurred only for normative dimensions useful for discriminating the ingroup from the outgroup. In a follow-up experiment, Belgian students were asked to imagine that, not British hooligans, but Belgian or German ones had caused the death of about forty people and the injuries of a few hundred during the Heysel tragedy in May 1985. Results again showed that ingroup members (Belgian hooligans) were rated more negatively than outgroup members (German hooligans). These results support the hypothesis that derogation of 'atypical' ingroup members can be concomitant with ingroup bias to the extent that it preserves the positive value assigned to the ingroup as a whole.

Deprez and Persoons' (1984) findings on the ethnolinguistic identity of Flemings in Brussels also manifest some kind of black sheep effect, even though the research was carried out without knowledge of Marques' experiments. In five different conditions, Deprez and Persoons investigated the attitudes of Flemish high school students in Brussels towards eight types of Brussels Flemings. These eight types differ systematically in their use of the French (outgroup's) language; three types are 'consistent' Flemings who do not speak French at home or at work, two types are 'adapting' Flemings who do not speak French at home but who speak French with their colleagues, and the other three types are 'Gallicised' Flemings who educate their children in French. The five experimental conditions were created by varying the characteristics of the investigator: whether he belonged to the Flemish ingroup or to the Francophone outgroup and the degree of his linguistic conformity with his audience; whether he conformed with it by speaking Dutch fluently or not, or diverged from it by speaking French. Results were extremely clear-cut. In the divergent conditions, subjects differentiated much more between the different types of Flemings: They devaluated 'Gallicised' Flemings and judged 'consistent' ones more favorably when confronted with an investigator belonging to an outgroup, as compared to the conditions where the investigator belonged to the ingroup or used the ingroup language.

LEADERSHIP AND WITHIN GROUP STATUS DIFFERENCES
IN INTERGROUP RELATIONS

Another finding of Sherif's experiments is also worth comment: Intergroup competition enhances the salience of the ingroup's hierarchical structures. This seems to be a more general effect of intergroup contact.

Harvey (1956) investigated groups of friends in a nursing school, using a sociometric questionnaire and other sources of information. The members of these groups were asked to write down, within a set time limit, as many names of towns as possible while listening to a record. They then watched a rather blurred projection of each list, and were asked to estimate the number of names of towns on each list. As predicted, there was a positive correlation between sociometric status and overestimation of attributed achievement. The groups then repeated the experimental task, some in the presence of a hostile group, the others in the presence of a friendly group. In this second session, both overestimation and its correlation with sociometric status increased only in the presence of a hostile group.

Mere reference to a rival group has an effect upon the pattern of influence within a group. In order to study the effect of invoking a rival group on the self-image of a membership group, we asked students at a school of architecture to give their opinions of their school, first individual opinions and then, after discussion, common opinions in groups of four, before finally again giving individual opinions. In a control condition, no reference was made to another school. In a second condition (the experimental condition), the subjects were also asked to estimate, in each case, the opinion likely to be expressed by the pupils of another school of high standing about the subjects' school (Doise, 1969). The effects of referring to the other group were various. The image of the subjects' own group became more clearly defined in the experimental condition. Subjects adhered more frequently to their group norms, and the extremists, i.e. those subjects who had expressed their opinions more clearly, enjoyed higher sociometric status and had more influence than was the case when the other school was not referred to. Mere reference to another group results, therefore, in one's opinions concerning one's own group becoming more extreme while, at the same time, the members of the group give special status to the most extreme among them.

Rabbie (1982) also reported that, during real interaction between groups, hierarchical roles seemed to be more easily differentiated when the interaction was competitive than when it was cooperative. On the other hand, group leaders could also use intergroup competition to ensure or reinforce their status in the group. The degree of competition induced by the leader was therefore studied, as a

dependent variable, as a function of stability of his status (Rabbie &
Bekkers, 1976). Leaders who could be deposed by a majority vote by
members of the group had a greater tendency to introduce competition between groups, especially when members of their group were
divided, and when their group occupied a comparatively strong
position relative to the other group. However, when their status was
more strongly threatened, they nevertheless tried to introduce intergroup competition, even when they had little chance of winning.
Rabbie and Bekkers (1976: 282) reported the results of a field study
which attempted to confirm these experimental results. In 29 Dutch
trade unions, members of the ruling committee and grass roots militants were questioned as to their degree of satisfaction with the way
in which their respective leaders conducted the affairs of the union.
A positive correlation was found between their discontent with the
leader and hostility towards opponents of the union, as expressed
by the leader in his speeches to the militants. Experimental research
has shown that it is the discontent of the members which could be
the source of the leader's aggressiveness towards opposing groups.
This would imply that a stronger within group conflict can result in
intergroup conflict aroused by threatened leaders.


THE CATEGORY DIFFERENTIATION PROCESS

The accentuation of contrasts between stimuli belonging to different
categories can be considered as a basic aspect of human information
processing. Certain Gestalt theorists (see Holzkamp, 1973), but especially Campbell (1956) and Tajfel and Wilkes (1963), have studied
characteristics of the processes. Quantitative differences between
stimuli belonging to two different categories are accentuated when
the category division is systematically related to the aspects being
judged. To give an example taken from Tajfel and Wilkes (1963), differences in size between lines belonging to a category A and lines
belonging to a category B are accentuated when all shorter lines
belong to one of those categories and all longer lines belong to the
other category. Such judgemental accentuations do not appear when
longer and shorter lines are randomly divided between the two categories. Tajfel (1959, Tajfel, Sheikh & Gardner, 1964) directly applied
the accentuation of contrasts model to judgements on resemblances
and differences between ethnic groups.

Accentuation of similarities of stimuli belonging to the same category are also predicted by the model, but, in general, such accentuation has not been as strong in judgements of physical stimuli (e.g. Tajfel & Wilkes, 1963) and of attitude statements (e.g. Eiser, 1971, see also Eiser & Van der Pligt, 1984). Furthermore, in judgements of people, similarities among outgroup members are more easily enhanced than similarities among ingroup members (Park & Rothbart, 1982).

There is also solid evidence that crossing two systems of categorization considerably weakens the category differentiation effect. This has been shown in several of Deschamps' experiments (Deschamps, 1977, Deschamps & Doise, 1978), which illustrate the effect of crossing categories. Arcuri (1982) and Vanbeselaere (1987) have replicated these findings. Brown and Turner (1979) have also replicated them in experimental conditions which involve all four groups resulting from crossing two categorization systems, but they did not observe less differentiation in conditions with only two of these four groups. The latter conditions correspond, in fact, to the usual dichotomic categorization where subjects are divided according to one criterion while sharing many other characteristics.

Categorization processes are not only judgemental processes, but they also manifest themselves in evaluations and behaviors. Judgemental, evaluative and behavioral differentiations are linked together in complex ways. Analyses on the societal level for instance show that differentiations and conflicts within a group can nevertheless maintain the overall cohesion of the group. This has already been hypothesized by Ross (1920) in his Principles of Sociology:

"The chief oppositions which occur in society are between individuals, sexes, ages, races, nationalities, sections, classes, political parties and religious sects. Several such oppositions may be in full swing at the same time, but the more numerous they are, the less menacing is any one. Every species of conflict interferes with every other species in society at the same time, save only when their lines of cleavage coincide; in which case they reinforce one another.
These various oppositions in society are like different wave series set up on opposite sides of a lake, which neutralize each other if the crests of one meet the troughs of the other, but which re-inforce each other if crest meets crest while trough meets trough.
A society, therefore, which is riven by a dozen oppositions along

lines running in every direction, may actually be in less danger of being torn with violence or falling to pieces than one split along just one line. For each new cleavage contributes to narrow the cross clefts, so that one might say that society is sewn together by its inner conflicts. It is not such a paradox after all if one remembers that every species of collective strife tends to knit together with a sense of fellowship the contenders on either side" (Ross, 1920: 164-165).

Lorwin (1972) has applied such an analysis to linguistic and ideological conflicts in Belgium and, according to him, it is the criss-crossing of these conflicts that has enabled that state to survive. Ethnological observations show that a cross-cutting structure, based on multiple memberships which hold across one another's boundaries, reduces the confrontations between the constituent parts of a society. It can readily be seen how marriage conventions can create one or the other type of society. A class which, in Jaulin's terminology (1973), refers to itself as 'one's own people' and consists of all those people like himself with whom sexual relations are necessarily incestuous, may become crossed with a class of 'other people' from which the individual may take a wife. Such a crossing of two groups may come about through cohabitation in collective houses, and represents a factor making for equilibrium and harmony: "If the basic social unit is a group in which one's own people and other people are bound together and 'intermixed', the other cannot be thought of negatively as non-self, but must be thought of first in a positive way, as being complementary." (Jaulin, 1973: 306). It is therefore possible that, according to marriage regulations, people may regard each other as belonging to different groups while at the same time they consider themselves as members of the same residential group.

The category differentiation process, as far as it can predict enhancement of similarities within groups and differences between groups, offers a firm cognitive basis for the dynamics postulated in the Sumner hypothesis. However, it seems in reality more complex, since, in intergroup relations, category differentiation functions asymmetrically and enhancement of intergroup similarities does not easily occur. On the other hand, the crossing of categories allows for homogeneity to be manifest both within and across group boundaries, or heterogeneity to be of the same importance in judging within or between group differences.

SOCIAL IDENTITY THEORY REVISITED

The assumptions of Henri Tajfel's social identity theory are now wellknown. An important aspect of an individual's identity results from his or her membership in social categories. Accentuating the differences between one's own social category and other relevant categories is a powerful source of positive self-evaluation. Another tenet of the social identity theory is the interpersonal-intergroup continuum: "What is meant by 'purely' interpersonal is any social encounter between two or more people in which all the interaction that takes place is determined by the personal relationships between the individuals and by their respective individual characteristics. The 'intergroup' extreme is that in which all of behavior of two or more individuals towards each other is determined by their membership of different social groups or categories." (Tajfel, 1981: 240).

More recently, Turner (1987: 45) developed a hierarchical model of identities by recognizing three levels of self-categorization:

"a)  the *superordinate* level of the self as a human-being;
 b)  the *intermediate* level of ingroup-outgroup categorization; and
 c) the *subordinate* level of personal self-categorizations based on differentiations between oneself as a unique individual and other ingroup members..."

Turner also postulates, according to Tajfel's continuum idea, "a functional antagonism between the salience of one level of self-categorization and other levels" (Turner, 1987: 49). This means, for instance, that focusing attention on individuals and their differences is negatively related to the perception of intra-class homogeneity or that, on the contrary, judgements in terms of common category membership are antagonistic to more personally oriented approaches. From this assumption follows that "factors which enhance the salience of ingroup-outgroup categorizations tend to increase the perceived identity (similarity, equivalence, interchangeability) between self and ingroup members (and difference from outgroup members) and so *depersonalize individual self-perception* on the stereotypical dimensions which define the relevant ingroup membership. *Depersonalization* refers to the process of 'self-stereotyping' whereby people come to perceive themselves more as the interchangeable exemplars of a social category than as unique personalities defined by their individual differences from others" (Turner, 1987: 50). These analyses, fitting with the continuum hypothesis, are undoubtedly inspired by a purely cognitive model postulating a kind of incompatibility between the attention directed to global or more particular features of objects in processing individual social cognitions (for an experimental study of similar processes in perception see Navon, 1977). Paradoxically, the cognitive processing model of Turner also prevents him from according a special status to the self as such, although his theory is called a 'self-categorization' theory.


INGROUP AND OUTGROUP DIFFERENTIATION

Another cognitive approach takes into account that information on one's own group members is treated differently than information on outgroup members. Such an approach was developed by Park & Rothbart (1982) and Quattrone (1986) when they studied perception of variability within ingroups and outgroups. Different levels of categorization are applied by individuals when encoding information about ingroup and outgroup members. Empirical results show that intragroup variability is often more accentuated than outgroup variability; the authors argue that this phenomenon is due to the different kinds of information individuals store and process when perceiving and judging ingroup and outgroup members: "The categories used to encode outgroup behavior are superordinate, general, and undifferentiated, whereas the categories used to encode ingroup behavior include more subordinate, differentiated categories as well. This implies that we learn the superordinate information about both ingroup and outgroup members, but we also learn more differentiated information about ingroup members" (Park & Rothbart, 1982: 1064). Consequently, individuals adopt different standards matching different levels of categorization in their judgments of ingroup and outgroup members. Theoretically, this process seems unrelated to the ethnocentric bias (Sumner, 1906), since it should occur as well with positive as with negative group characteristics.

Experiments in more dynamic settings, carried out with 'natural' (Lalonde, Moghaddam & Taylor, 1987) as well as with *ad hoc* groups (Worchel, 1987), also show more complicated links between accentuation of within group similarity and between group differentiation. Intragroup differentiations do not necessarily disappear when intergroup differentiations are made salient (Moghaddam & Stringer, 1986; Taylor & Moghaddam, 1987). In experiments, most participating groups are at an early stage of development. That may explain the more frequently found intragroup homogeneity. Worchel (1987) argues in favour of such a dynamic pattern of differentiation: members of groups that are in an early stage of establishing an identity will have a tendency to perceive greater ingroup than outgroup homogeneity; however, after identity has been established, members may focus on the diversity that is present within their group.

A detailed inspection of earlier results obtained by Turner (1975), in an experiment including two stages, also shows that ingroup differentiation can occur at the same time as between group differentiation, and that these differentiations are therefore not to be considered as antagonistic processes. In some of his experimental conditions, prior decisions had to be made, as in the classical Tajfel (1970) paradigm, for an anonymous ingroup member and an outgroup member in order to generate intergroup differentiation. This differentiation occurred, but subsequent choices for self and other ingroup members remained systematically biased in favour of the self (Turner, 1975, tables 3 and 4), as much as, or even more than, in a situation without previous intergroup differentiation. Contrary to Turner's ideas, differentiation at the 'intermediate' (ingroup-outgroup) level of categorization does not seem inversely related to or incompatible with differentiation at the lower categorization level of individuals as such.

Such findings demand models that integrate simultaneous differentiations between groups and within groups. Prototypical categorization models (e.g. Rosch, 1978) can be used for such purposes because a prototype is defined concomitantly in terms of differences between other elements of its membership category (fully represented by the prototype), as well as in terms of differences from elements of another category. In this sense, prototypical categorization could simultaneously account for both within category singularization and between category differentiation.

For analyzing categorization processes within the category of French executives, Boltanski (1982) based his interpretation upon Eleonor Rosch's (1978) prototypical categorization model. Executives are a very heterogeneous group, which is shown in the representation they construct of themselves. Boltanski's analysis goes beyond a mere description by indicating the functions assumed by these not very clear-cut categories, for example: to generate fierce competition in situations which cannot be precisely evaluated due to the variety of status indices and symbols; to mask significant lines of cleavage in the group through criss-crossing of differentiated memberships; to facilitate inclusions in and exclusions from the membership group according to specific circumstances and strategies. As such, one can paradoxically speak of "the cohesion of fuzzy sets" or of "the strength of weak aggregates".

According to current categorization models (Cantor & Mischel, 1979), the definition of more inclusive classes is subject to less variation and involves a smaller number of distinctive traits than the definition of more particular classes. However, several studies (e.g. Rosch et al., 1976; Deaux et al., 1985) have failed to detect differences in the number and content of attributes individuals associate with categories at different levels of abstraction. It then seems plausible to expect comparable degrees of between subjects variability in perceptions at different levels of social categorization. In a study using a classical stereotypes questionnaire, such results were found (Doise & Lorenzi-Cioldi, 1987). About 500 subjects (Swiss and foreign high school students in Geneva) were asked to rate several targets ("yourself, the Swiss, the Foreigners, your friends, and people of your nationality") on a trait list adapted from Peabody (1985). Between subjects variability for each target's judgements, across the traits, were very similar. That is to say, between subjects variability was not more important in the perception of single and concrete targets (such as 'yourself' or 'your friends') than in the perception of a more collective entity (such as the Swiss).

COMPARISON DIMENSIONS

Another caveat against the assumption of a negative relationship between differentiations at different hierarchical categorization levels stems from the consideration that the nature and content of comparison dimensions can affect cognitive processes: performance or ability dimensions enhance comparisons between individuals who seek interpersonal distinctiveness, and emotional or affective dimensions tend to enhance interpersonal cohesion, in which individuals seek similarity (Leventhal, 1970; Tesser & Campbell, 1983). A large proportion of empirical research reveals that such 'agentic' and 'relational' dimensions are matched with stereotypes built around particular social groups, especially around groups occupying different positions in the social structure (for instance, ethnic groups: Marin & Triandis, 1985; gender groups: Snoodgrass, 1985).

More generally, members of high status groups appear to bolster interpersonal distinctiveness on 'agentic' dimensions, whereas members of low status groups strengthen a sense of group cohesion or fusion on more 'relational' dimensions (Lorenzi-Cioldi, 1988). More complicated patterns of differentiation arise when we consider relative status of groups in interaction. It seems likely that the search for interpersonal distinctiveness depends on the individuals' positions in a network of intergroup relations. Members of dominating groups consider themselves individually as the point of reference in relation to which other people are defined: They perceive themselves as unique individuals and do not seek self definition in terms of group membership. On the other hand, members of dominated groups define themselves, and are also defined by others, more in terms of social categorizations imposed on them. The search for differentiation from other individuals would therefore be stronger for members of dominant groups and would be enhanced when such group affiliation is made salient. This has led Lorenzi-Cioldi (1988) to propose a social categorization approach which takes into account social asymmetries and distinguishes between different sorts of groups: On the one hand, there would be dominant or 'collection' groups; social categories made up of individuals perceiving themselves as distinct one from another; on the other hand, there would be dominated or 'aggregate' groups, made up of individuals defining themselves primarily in holistic terms which distinguish their group from other groups, while stressing the similarity of ingroup members more similar to one another at the personal level.

Some empirical results uphold these propositions. Within the framework of Tajfel's social identity theory, Brown and Ross (1982) reported results of a factor analysis summarizing data obtained in an experimental intergroup setting involving different status groups; these results "showed a clear negative correlation between the 'individualist' factor identified as 'social mobility' and the 'collectivist' factor of positive intergroup differentiation", i.e. an inverse relationship between concerns about personal and collective interests. However, the authors also reported "results from the perceived homogeneity measure which proved unrelated to either of these factors and relatively insensitive to the experimental manipulations." (Brown & Ross, 1982:175). Brown and Williams (1984) tested the reverse hypothesis that an increase in ingroup identification should be linked with intergroup differentiation. Data on several groups of workers in a factory were collected, but the predicted link was found primarily in groups with a relatively low status in the organizational structure. In our study, mentioned above (Doise & Lorenzi-Cioldi, 1987), groups could be classified according to the prestige of their national citizenship in the context of Swiss society (Swiss nationals versus Spanish or Italians). The index of variability of the collective stereotypes ('the Swiss' and 'the Foreigners') show that, regardless of subjects' national origins, judgements about 'the Swiss' displayed more between subjects variance than judgements about 'the Foreigners'. That is to say, there seems to be less agreement, in certain circumstances, in defining the characteristics of a privileged group than those of a less privileged one.

CONCLUSION

Social psychological studies of intergroup relations have made a great effort to experimentally illustrate Sumner's (1906) classical thesis, but more careful inspection of these experimental findings shows that the correlation between ingroup solidarity and outgroup antagonism is not always as strong as is often taken for granted. Intergroup differentiation and, more specifically intergroup hostility can also generate intragroup differentiation and conflict. Especially when relations between self and other ingroup members are brought into focus, correlations appear which are contrary to Sumner's hypothesis.

Ideas on covariation between ingroup and outgroup differentiation have recently been introduced in the social-psychological study of intergroup relations.

From this, we do not draw the conclusion that traditional categorization processes (in the sense of Tajfel) bearing on intergroup differentiation and intragroup homogenization have to be abandoned in favor of more recent models or conjectures. It is undoubtedly true that, in some situations, within group and between group differentiation is inversely related; on the other hand, direct links have nevertheless been conceptualized. Relative status positions of different ethnic groups and comparison dimensions salient in interethnic encounters particularly may further specific kinds of ingroup differentiation which are compatible with between group discrimination and conflict. Researchers usually argue in favor of their own model, but we have reached a point where different models seem well established and where there is a urgent need to attempt to integrate or articulate different models (see Doise, 1986). Several categorization models are now available, and the main aim of research should now be to determine under which conditions they apply. This is especially required when the issues at stake are socially relevant, as is often the case with problems pertaining to intergroup relations.

# 4

# STRATEGIES OF
# IDENTITY MANAGEMENT

Ad van Knippenberg
Catholic University Nijmegen
The Netherlands

In many, if not all, Western societies, there seems to be a basic tendency for individuals to try to improve their social position. There is little doubt that much of this upward striving has to do with purely material considerations, i.e. with fulfilling physical needs and providing for a more secure and comfortable life. However, it is now widely recognized in social psychology that the search for status and prestige is also intrinsically motivated. It is assumed that people prefer to think positively about themselves (e.g. Festinger, 1954), they like others to view them in a positive light (e.g. Schlenker, 1975) and they like the groups to which they belong to compare favorably to other groups (e.g. Tajfel, 1978).

Despite this general tendency towards upward locomotion, there appear to be large differences between individuals and between groups with regard to the manifest intensity of upward striving and the strategies employed to gain higher status and prestige. We assume that, to some extent, these differences are related to the objectively given and subjectively perceived possibilities for individuals and groups to get access to high status positions, prestigious jobs, or, more generally, to positive social recognition. For instance, ethnic minorities will probably encounter more difficulties in achieving and maintaining a positive social identity than members of the majority. My approach to this issue will be one in which identity management problems of minority groups are treated in general, theoretical terms rather than one which focuses on particular minority phenomena in our society.

In this chapter, I will discuss some aspects of the relationship between social conditions (i.e. socio-structural variables like permeability of group boundaries, stability of intergroup relations and relative size, status and power of the groups involved) on the one hand and the strategies of identity management to which individual group members resort on the other hand. We start our discussion of socio-structural conditions and some of their psychological consequences with an early study of how the prospect of social mobility affects an individual's satisfaction, together with some of the initial theorizing aimed at explaining it. Then a brief outline of social identity theory is presented, in which the significance of group status is emphasized. This is followed by a discussion of the implications of the permeability of group boundaries for the strategies that are used for identity enhancement. In the next section, a more systematic account of identity management strategies is presented. Subsequently, the social psychological effects of stable and unstable intergroup relations are analyzed. The chapter concludes with a discussion of some aspects of minority-majority relations in the light of the preceding theoretical considerations.

INDIVIDUAL MOBILITY AND RELATIVE DEPRIVATION

In *The American Soldier*, Stouffer, Suchman, DeVinney, Star and Williams (1949) described an interesting phenomenon. In the United States army, soldiers serving in some units had more opportunities to pass to a higher rank than soldiers in other units. Unexpectedly, the soldiers who had better opportunities were *less* satisfied with their position than soldiers who had little prospect of getting a promotion. Merton and Kitt (1950) suggested the following theoretical explanation of the phenomenon described by Stouffer *et al*. They assumed that people choose a reference group in order to evaluate their own outcomes. Soldiers serving in units in which rapid advancements were possible, probably compared their own outcomes with others who had already improved their position. Because such a comparison would generally turn out to be unfavorable for the subject, they tended to be less satisfied with their own situation. In units in which practically no promotions took place, it is assumed that the soldiers compared themselves with their colleagues, who, like themselves, were not promoted. In general, such comparisons were not unfavorable and, therefore, these soldiers were more satisfied with their position.

The explanation of Merton and Kitt was later elaborated into the 'Relative Deprivation theory' (cf. Davis, 1959; Hyman & Singer, 1968; Gurr, 1970). In this theory, the term 'reference group' occupies a central place. Sherif and Sherif (1969) define reference group as a group with which an individual identifies because (s)he wants to become a member of this group or, if (s)he already is, to stay a member of it. People evaluate their own achievements or outcomes by comparison with those of their reference group. In general, people prefer reference groups with higher status (Martin, 1981). However, the choice of a comparison group also depends on the subjectively perceived possibility of gaining membership in the reference group; in other words, the choice of reference group also depends on the 'feasibility' of the aspired to position (cf. Crosby, 1976; Cook, Crosby & Hennigan, 1977).

This theoretical explanation focuses on the evaluation of individual outcomes by comparing them with a reference group standard or norm, which in turn is derived from what other individual members of the reference group (either own group or a group of which the individual wants to become a member) have achieved. Satisfaction with the individual outcome is thus essentially a function of interindividual comparisons. Although Runciman (1966) already pointed out that this type of interindividual ('egoistic') relative deprivation must be distinguished from relative deprivation resulting from intergroup comparisons (i.e. 'fraternal' relative deprivation), later studies have continued to explain various large-scale social phenomena (e.g. social discontent and unrest) predominantly on the basis of interindividual outcome comparisons (e.g. Gurr, 1970). In a comprehensive analysis of the effects of socio-structural conditions, the implications of intergroup comparisons need to be more fully incorporated (see Billig, 1976; Tajfel, 1978; Walker & Pettigrew, 1984). In my opinion, social identity theory is more suited for that purpose than relative deprivation theory because it is broader in scope. Social identity theory covers a much wider range of psychological and behavioral phenomena associated with group membership, status of the group and individual prospects for upward mobility. Therefore, I assume that an analysis from the point of view of social identity theory will provide a more systematic account of how group members act upon variations in socio-structural conditions.

## SOCIAL IDENTITY AND GROUP STATUS

In the late sixties and early seventies, Tajfel and co-workers developed the Social Identity Theory (Tajfel, 1972, 1978; Tajfel & Turner, 1979). I will first briefly introduce the main principles of the theory and subsequently elaborate some of its implications for our discussion of the effects of socio-structural conditions on identity management strategies.

Social identity was defined "as that *part* of an individual's self-concept which derives from his knowledge of his membership of a social group (or groups) together with the value and emotional significance attached to that membership" (Tajfel, 1978, p. 63). The relationship between self-concept and group membership was established by using social categorization theory. Because individuals are perceived and perceive themselves as belonging to social groups or categories (e.g. 'men', 'women', 'blacks', 'whites' etc.), and social categories tend to be associated with specific characteristics (e.g. loyalty, aggressiveness, zeal), individual category members may gradually incorporate such category traits as inherent features of their self-concept, or at least of their situationally determined self-definitions (cf. Turner, 1987).

Social identity theory further assumes that people strive for a *positively valued* social identity. Since the value of a social category is established through comparison with other relevant social categories, it follows from this assumption that individuals will try to differentiate their own group from relevant comparison groups in a positively valued direction.

Society consists of various interlocking category systems, e.g. of categorizations in terms of professions, job-categories, sex, socio-economic background or ethnic groups. These categorizations constitute status hierarchies, that is there exist rank orderings of social groups in terms of socially recognized value, worth or prestige. The higher a group is in the status hierarchy, the more this group can contribute to the positive social identity of its members. The lower the status position of the group, the more negative are the social identity consequences for its members.

With regard to the status hierarchies that exist in a given society, it seems useful to make a distinction between what may be tentatively called *direct* and *indirect* status hierarchies. In direct (or primary) status hierarchies (e.g. professions, jobs, public offices, sporting teams), group statuses reflect the socially dominant evaluation of the particular performances and achievements of these groups. In contrast, indirect (or secondary) status hierarchies (e.g. sex, ethnic groups, religious groups, regional background) are mainly socially significant to the extent that they affect the admission of individuals to groups in the direct status hierarchy (due to e.g. institutionalized barriers, explicit prohibition, social disapproval or self-selection). [1]

The positions of groups in direct status hierarchies reflect the dominant values in society; it shows how much regard people have for e.g. nursing as compared to teaching or cleaning or building houses or defending the country, etc. The status positions in an indirect hierarchy betray the implicit or explicit beliefs people have about the characteristics of specific social categories, whether the members of these groups possess the necessary qualities for a particular job, or, even if they have the qualities, whether it is 'appropriate' for them to engage in such-and-such activities. As we shall see later, the degree to which direct status groups are accessible to members of different indirect status groups is an important determinant of how people manage their identity enhancement needs.

GROUP STATUS AND PERMEABILITY OF GROUP BOUNDARIES

The status positions which groups occupy in society and the permeability of group boundaries, that is the extent to which individuals are free to join or leave these groups [2], may be considered to be two key variables of social identity theory. Following from the definition of social identity as that part of an individual's self-concept which derives from his membership of a social group, it may be argued that:

(a) It can be assumed that an individual will tend to remain a member of a group and seek membership of new groups if these groups have some contribution to make to the positive aspects of his social identity; [...]

(b) If a group does not satisfy this requirement, the individual will tend to leave it *unless:*

(i) leaving the group is impossible for some 'objective' reasons, or;

(ii) it conflicts with important values which are themselves a part of his acceptable self-image.

(c) If leaving the group presents the difficulties just mentioned, then at least two solutions are possible:

(i) to change one's interpretations of the attributes of the group so that its unwelcome features (e.g. low status) are either justified or made acceptable through a reinterpretation, or;

(ii) to accept the situation for what it is and engage in social action which would lead to desirable changes in the situation [...]

(d) No group lives alone – [...] the "positive aspects of social identity" and the reinterpretation of attributes and engagement in social action only acquire meaning in relation to, or in comparison with, other groups (Tajfel, 1978, p. 64).

As the above quotation suggests, the interaction of group status and permeability of group boundaries determines the strategies available for the enhancement of their members' social identity. Let us briefly discuss the main characteristics of this interaction.

High status groups contribute to their members' positive social identity while low status groups generally fail to do so. In line with this argument, research has indeed shown that membership in low status groups has a negative effect on the self-esteem of individuals (cf. Brown & Lohr, 1987; Wagner, Lampen & Sylwasschy, 1986) and that members of low status groups try to dissociate themselves psychologically from their group (cf. Jahoda, 1961; Klineberg & Zavalloni, 1969).

When group boundaries are permeable, the dominant strategy of social identity enhancement for low status group members is to join higher status groups. In an experimental study of a four-group hierarchy (Ross, (1975, cited in Tajfel & Turner, 1979), it was found that the desire to pass upwards into another group increased with decreasing group status. In research with regard to communication patterns (Cohen, 1958), evaluation of group characteristics (Van Knippenberg, 1978) and group membership aspirations (Mann, 1961), there appeared to be a tendency for members of low status groups who anticipate upward social mobility to focus their communications, evaluations and aspirations on the higher status group instead of on their own group. It is assumed that, in such situations, people identify with the higher status group before actually passing to it. Thus, it appears that, in situations in which a (low status) group cannot make a satisfactory contribution to the social identity of its members, individuals will try to leave this group in order to gain membership in a higher status group. Obviously, this identity enhancement strategy can only be used effectively by individual group members who possess whatever it takes (e.g. ability) to move upward into a higher status group.

When group boundaries are impermeable, members of low status groups will tend to enhance their social identity by improving the position of their present group as a whole. As outlined above, individual locomotion to a higher status group may be the preferred way to improve one's social identity. In some situations, however, changing group membership is virtually impossible, e.g. in a strongly segregated society. In such situations, that is when individual mobility is not feasible, ingroup identification may be inevitable. Under these circumstances, members of low status groups will have to find other ways to improve their social identity. One specific option would be to elevate the status of their group as a whole, for instance by competing with higher status groups. Some other strategies will be discussed later.

This discussion of the interaction between group status and permeability may be summarized by the contention that individuals prefer to maintain membership in high status groups or, if they are member of a lower status group with sufficient individual ability, they will seek membership in high status groups. If upward mobility is not possible, one salient option is to engage in intergroup competition to improve the relative position of the group as a whole.

The interactions of group status, permeability of group boundaries and individual ability were investigated in an experiment by Ellemers, Van Knippenberg, De Vries and Wilke (1987). In this experiment, subjects were randomly allocated to one of five three-person groups in the laboratory. After performing an individual task and 'the first round' of a group task, the subjects received bogus feedback about their individual ability compared to that of their fellow group members, and about the performance of their group. They were either told that their group performed well (high group status, i.e. the second best group) or poorly (low group status, i.e. the second worst group). It was further announced to the subjects that in later rounds of the group task they would all remain a member of their present group (impermeable boundaries), or that some of the participants might move from one group to another (permeable boundaries). The central dependent variable was *group identification*, measured by a set of nine questions (e.g. "Do you think that you have more in common with the members of your group than with members of other groups?")

The most important results with regard to ingroup identification were:

a Members of high status groups showed stronger ingroup identification than members of low status groups;

b In low status groups, permeable group boundaries invoked significantly lower ingroup identification than impermeable boundaries;

c In low status groups ingroup identification decreased as individual ability increased (while in high status groups there was no such relationship).

In addition, Ellemers *et al.* measured whether the subjects were satisfied with the division into groups and with the performance of their group. It is interesting to note that – while irrespective of permeability condition, members of low status group were less satisfied with their group's performance than members of high status groups – only in the *permeable* condition were members of low status groups less satisfied with the division into groups than members of high status groups; in the impermeable condition there was no effect of group status on satisfaction with the division into groups. These results indicate that impermeable boundaries help to reconcile subjects to their group membership (since it is inevitable), but they are not necessarily satisfied with the 'status' (i.e. the performance) of their group.

The results of the experiment of Ellemers *et al.* thus illustrate the importance of the permeability of group boundaries with regard to the identity enhancement strategies individuals resort to. The sort of situation studied, however, was of necessity quite limited, i.e. subjects were only allowed to increase or decrease their ingroup identification. A more systematic analysis of the strategies people may use to manage their social identity seems necessary.

STRATEGIES OF IDENTITY MANAGEMENT

Various authors have put forward suggestions with regard to the ways in which people may protect or enhance their (social) identity (see, for instance, Turner & Brown, 1978; Rijsman, 1980; Van Knippenberg, 1984). Although the labels used tend to vary, there is substantial correspondence between the suggested options. Below, a list of options is presented, including some alternatives which have not been mentioned elsewhere.

1  The most prominent strategy, in my view, is to join high status groups (upward mobility) and to maintain membership in high status groups.
2  If one is a member of a low status group, one may improve one's social identity by improving the relative status position of the ingroup as a whole:
    (2a) by competing with high status groups in terms of established criteria of ranking. This strategy may be labeled social competition (Turner, 1975).
    (2b) by contesting the validity or legitimacy of existing criteria for status allocation (or at least the exclusive application of these criteria), to propose alternative criteria on which the ingroup occupies a better position, and to try to gain social acceptance for these alternative criteria (cf. Lemaine, 1974).

It should be noted that intergroup comparisons take place in a multidimensional comparison situation: comparisons are not only made in terms of prestige, income or intellectual achievements, but also focus on moral or religious matters, on the merits of cultural achievements and even on aesthetic aspects of, e.g. language and appearance (e.g. 'Black is beautiful'). In the context of such a multidimensional comparison situation, groups may sometimes concede outgroup superiority in important areas, while simultaneously claiming ingroup superiority in other areas which are also important. In fact, there is a definite tendency to find ingroup assets more important than those of the outgroup (cf. Van Knippenberg, 1978), a tendency for which Mummenday and Schreiber (1984) coined the expression: 'different but better'. It is plausible that the processes outlined here may operate as an intergroup role differentiation process in which different groups fulfil complementary functions to reach a common goal, which might render stability and some degree of positivity to the intergroup situation (cf. Rijsman, 1980; Van Knippenberg, 1978; Brown & Wade, 1987). Because of these implications, this strategy may be labeled 'social cooperation' (cf. Van Knippenberg, 1984).

3 In situations in which low status groups cannot possibly improve their position (and, as will be argued later, high status groups have a secure position), group members may cease to make intergroup comparisons (at least these high vs. low status group comparisons). In these circumstances, (social) identity is established through comparisons between finer categorical distinctions or through interindividual comparisons.

4 Group members in low status groups may improve their identity by making other categorical differentiations more salient, at least when this re-arrangement puts them in a more favorable position. For instance, persons may prefer to categorize themselves as supporters of a certain football club rather than e.g. as 'unemployed'. A concomitant line of action would be to de-emphasize unfavorable category membership (e.g. decrease its visibility or distinctiveness).

5 Finally, there is a broad cluster of options which are all more or less dysfunctional or even pathological, e.g. denial, withdrawal, fantasy comparisons, idiosyncratic perceptions of reality, self-denigration or self-hate, (cf. Lewin, 1948) etc.

For each of these strategical options, real life examples and, sometimes, illustrations from social psychological research may be found. Within the framework of this chapter, however, only the first three strategies will be further elaborated in relation to another important socio-structural variable, i.e. the stability of the intergroup relations. As I shall argue below, identity management strategies are a function of both the stability of the intergroup relations and the permeability of group boundaries.

THE CASE OF STABLE INTERGROUP RELATIONS

A status relationship between two groups is considered to be stable if one believes that it cannot change. It is often assumed that such a situation of social quiescence exists only in theory, or in the somewhat romantic examples of disappearing non-western or ancient cultures, in which intergroup relations simply did not change for ages. However, stable intergroup relations may not be quite that rare.

In order to analyse the effects of stable intergroup relations for the social identities of the people living in such a system, a distinction must be made between two types of relatively stable intergroup situations which are briefly denoted here as 'rigid caste system' and 'dynamic equilibrium', respectively. Let us look at each of these systems in some detail.

## The rigid caste system

This type of system comprises stable intergroup situations, perceived as highly legitimate and supported by deeply entrenched social beliefs that the intergroup relations reflect the natural order of things. In these situations, the intergroup status differences are firmly rooted in belief structures containing notions about a 'natural' social stratification in which one group is 'inherently' of higher status than the other. Furthermore, in these rigid caste systems, group boundaries are highly impermeable: individuals cannot possibly move from one group to the other (i.e. neither in upward nor in downward direction). Individuals are born into their respective groups and the characteristics associated with high and low status are thought to be hereditary. Examples of these kinds of belief structures may be found in many historic cultures, e.g. in the feudal system in medieval Europe, where sharp and impermeable distinctions separated nobility from predials and serfs. Some of these notions persisted in later ages, in which inherited differences were still often attributed to higher and lower classes. Many social philosophies explicitly incorporated hereditary concepts to justify intergroup status differences and various kinds of social practices used to maintain the status quo (see, for instance, Hughes, 1986).

As was pointed out by Tajfel (1978), the existence of basic 'constitutional' or hereditary differences between groups may prevent the members of different status groups from being conceived of as the same sort of people, thereby eliminating the possibility of intergroup comparisons. Intergroup comparisons require that the social categories involved can be grouped under a common denominator at a higher level of categorization (cf. Turner, 1987). In this way, notions concerning hereditary intergroup differences serve to sustain the stability of the intergroup status relations, because by suppressing intergroup comparisons, lower status groups will not interpret the situation as one in which they are relatively (fraternally) deprived.

In view of these considerations, it may be argued that, in situations of stable intergroup status differences of the rigid type, the social identity of high and low status groups cannnot be established by mutual comparison. In terms of identity management strategies, positive self-definitions can only be realized through intragroup comparison and by differentiation in terms of finer social categorizations, i.e. by differentiating one's own subgroup from relevant other subgroups (within the general – lower or higher – social class to which one belongs).

## The dynamic equilibrium

A second type of relatively stable intergroup relations may be found in a fairly open society in which group boundaries are highly permeable. In our view, the intergroup status rank order (e.g., comprising public offices, industry, commerce, education etc.) may be largely maintained over time in societies in which a high degree of permeability of group boundaries is achieved. We assume that a high level of permeability, on the basis of what members of a society perceive as reasonable and just criteria, serves to legitimize the existing intergroup status relations. Obvious candidates for reasonable and just criteria for access to social groups in the direct status hierarchies (e.g. professional categories, jobs, sporting teams) seem to be ability and effort (cf. Taylor & McKirnan, 1984). In our present argument, it is assumed that the status positions of groups in such a system are kept in place by cybernetic mechanisms. For instance, the temporary decline of the social prestige of, say, teaching would probably result in a reduced interest in this professional activity up to a point where the status of this group would have to be reconsidered in order to restore the balance.

The most salient identity enhancement strategy in a dynamic inter-group status equilibrium is individual mobility: members of a lower status group will try to improve their position by joining a higher status group. Lower status group members who lack the necessary talents, or fail to invest the required effort, will not be able to mobilize forces in order to change the group status as a whole, because (1) relatively competent group members tend to pass to higher status groups, which would handicap the group in an open competition and (2) there would be little social support for a reconsideration of the established status criteria, because permeable group boundaries will have rendered legitimacy to the existing criteria. Those who stay behind may have to resort to intra-group comparisons (or focus comparisons on still lower status groups; e.g. 'poor-white racism') in order to achieve a satisfactory social identity.

## THE CASE OF UNSTABLE INTERGROUP RELATIONS

There may be numerous causes for the transition of stable inter-group relations to unstable ones. A full account, or a systematic taxonomy of such causes, is beyond the scope of this chapter. Therefore, I will confine myself to mentioning a few suggestive examples, e.g. the development of new technologies, the emergence of new sources of wealth and power, the death of a charismatic leader, the decline of faith in the system as a result of corruption and inefficiency, the gradual decay of system maintenance mechanisms, etc. Roughly, two broad categories of causes of instability may be distinguished: changes in the power equilibrium and decreasing legitimacy of the system. These two processes may, of course, also reinforce each other.

For the present discussion, the effects of instability are more important than the causes. On the basis of theoretical, as well as empirical, considerations, Van Knippenberg (1984) argued that the effect of increased instability is, in general, to raise the salience of the inter-group comparison for both low and high status groups.

For low status groups, the awareness that the status relations between their own group and the higher status group may be changeable, even as a fairly remote possibility, may be accompanied by a rapidly increasing salience of that particular intergroup comparison (particularly since, both in interpersonal and in intergroup relations, there seems to be a basic tendency to focus on upward comparisons, cf. Wheeler, 1966; Van Knippenberg, 1978). Moreover, from the perspective of the relatively lower status group, unstable intergroup relations also tend to coincide with perceived illegitimacy of the status relationship. To be sure, the directions of causality between the variables (in)stability, illegitimacy and comparison salience cannot be easily established, although, at least theoretically, increased comparison salience may be argued to follow from instability and perceived illegitimacy (cf. Tajfel, 1978; Tajfel & Turner, 1979). In turn, increased comparison salience is predicted to enhance intergroup differentiation tendencies for low as well as high status groups (see Doise & Sinclair, 1973; Van Knippenberg, 1978; Van Knippenberg & Wilke, 1979).

In general, members of high status groups are predicted to feel little need for downward comparisons. At least in situations of social quiescence, comparisons with lower status groups will not be salient. When, however, as a result of increased instability, the ingroup's superiority is threatened, measures have to be taken to secure its position. Increased instability will lead to an enhanced concern with the intergroup relationship. It sensitizes the higher status group to downward intergroup comparisons, it makes this comparison more salient and, therefore, leads to increased efforts to re-establish a comfortable positive intergroup distinctiveness.

MINORITIES AND MAJORITIES: GROUP SIZE, STATUS AND POWER

The terms *minority* group and *ethnic minority*, as they are used in everyday language, cannot easily be translated in simple unidimensional social psychological concepts. The word minority suggests relatively small group size, but it also has additional connotations. In practice, minorities tend to have lower status and less power than majorities. Generally, the term minority is used to designate groups who are, at least to some extent, physically distinct (e.g. in appearance or in speech), which enables the social environment (the 'majority') to react differentially or in a discriminatory way to members of minority groups.

In brief, a cluster of potentially important variables may play a role in the social psychology of minority groups. In my view, social psychology has hardly begun to disentangle the interplay between these socio-structural variables with regard to their effects on identity management strategies of minority groups. Below, I will first present some tentative conclusions with regard to the effect of relative group size, status and power. In the final section, the implications of our theoretical analyses for identity enhancement strategies of minority groups will be discussed.

An important finding supporting social identity theory was that 'minimal groups' displayed ingroup favoritism. Groups formed in the laboratory, on a trivial or random basis, without interaction between group members, showed ingroup favoritism, i.e. individuals allocated more money (or points) to anonymous fellow ingroup members than to anonymous outgroup members, although the subjects themselves did not personally benefit from their own money allocations (e.g. Tajfel, 1970; Tajfel, Billig, Bundy & Flament, 1971; Billig & Tajfel, 1973). These findings were interpreted as resulting from a tendency to create positive intergroup distinctiveness.

Earlier research, using different paradigms, suggests that minority groups might not show such ingroup favoritism. For instance, Clark and Clark (1947) found that black children tended to prefer white dolls to black ones. Vaughan (1964) found that Maori children in New Zealand evaluated their own group, on several measures, less positively than the dominant white group.

In further research on intergroup differentiation between minorities and majorities, the results were contradictory. Gerard and Hoyt (1974) found that two-person minority groups displayed more ingroup favoritism than five-person (equal) and eight-person (majority) groups. Sachdev and Bourhis (1984) found that experimental minorities were less fair and showed more ingroup favoritism than majorities. Branthwaite and Jones (1975) observed that Welsh subjects in Cardiff showed more ingroup favoritism than English subjects. Vaughan (1978) replicated his earlier (1964) study and found that, ten years later, Maori subjects in urban areas showed as much ingroup favoritism as white ('Pakeha') subjects.

Experimentally manipulated differences in power and status also affect intergroup differentiation. Sachdev and Bourhis (1985) found that high power groups showed much stronger ingroup favoritism than low power groups. Sachdev and Bourhis (1987) observed more discriminatory behavior in high and equal status groups than in low status groups. A similar tendency was reported by Ng (1986).

A tentative solution for these apparent empirical contradictions may be found in a suggestion put forward by Moscovici and Paicheler (1978). These authors argued that ingroup favoritism and outgroup favoritism may have different meanings for minorities and majorities in different social conditions. Elaborating their ideas in our own terms, distinctions need to be made between:

1 *Minorities in insecure positions.* These minorities, who lack a binding assertive ingroup ideology, will probably show outgroup favoritism (that is, vis-à-vis the majority). Examples are provided in the studies of Clark and Clark (1947) and Vaughan (1964).

2 *Self-confident, assertive minorities.* Such minorities, who possess a clear purpose and a consistent ideology, will tend to show ingroup favoritism. Examples given by Moscovici and Paicheler include the Zionist movement for Jews and the Black Power movement in the USA. The Welsh subjects in the Branthwaite and Jones (1975) study and the urban Maori subjects in Vaughan's (1978) study also constitute examples of this phenomenon.

3 *Insecure, threatened majorities.* In these conditions majorities will show ingroup favoritism (vis-à-vis minorities). As noted earlier, unstable intergroup relations may evoke defensive reactions from high status groups. Also, majorities who have difficulty in defining a positively valued social identity may display strong ingroup favoritism.

4 *Secure majorities.* In stable intergroup relations, or given large intergroup differences, the privileged majority can afford to benevolently concede advantages to non-threatening minority outgroups.

As these propositions (for which Moscovici and Paicheler provided experimental evidence) suggest, relative ingroup size and stability of intergroup status relationships (which is assumed to be related to assertiveness of minorities and (in)security of majorities) seem, in interaction, to affect intergroup differentiation behavior. Needless to say, further theoretical elaborations and empirical research are required in order to reach a more complete understanding of the interactions between the variables involved.

STAGES IN MINORITY-MAJORITY RELATIONS AND IDENTITY MANAGEMENT STRATEGIES

Above, a distinction was made between two broad categories of minority group attitudes and the corresponding, somewhat complementary, majority group reactions. In this final section, these observations will be related to the foregoing theoretical analysis of the effects of socio-structural variables like stability of intergroup status relationships and permeability of group boundaries.

Within contemporary western societies, as well as in the history of western cultures, a great variety of ingroup-outgroup relations may be observed. As Taylor and McKirnan (1984) propose, it is possible to distinguish different stages in intergroup relations, which tend to evolve in chronological sequence from one stage to the other. In contrast to the five-stage model of Taylor and McKirnan, the present proposition contains three, more basic stages, viz.:

1 Rigid, stable intergroup relations without intergroup mobility. In the dominant (and virtually undisputed) social representations, intergroup status differences are attributed to inherited differences. In this stage, intergroup (minority-majority) comparisons are suppressed. If comparisons were invoked explicitly, minority groups would show outgroup favoritism (i.e. conforming to dominant evaluations). The attitudes of insecure minorities outlined in the previous section may be understood as remnants of this general class of minority-majority social representations. Finer categorizations, intragroup comparisons and various kinds of dysfunctional identity management strategies may be observed in these situations.

2 Unstable intergroup relations in which various minority groups gradually develop positive ingroup identities and challenge the legitimacy of the dominant views. Both social competition and social cooperation strategies may be used as a function of utilitarian considerations. In explicit intergroup comparisons, minority groups show strong ingroup favoritism. Collective actions to enhance positive ingroup identity (e.g. emancipation movements)

serve to counteract persisting negative group images and, ultimately, aim at removing stereotypical normative and institutional barriers blocking or inhibiting upward social mobility of members of minority groups.

3 Stable group statuses in a dynamic equilibrium with a high level of permeability of group boundaries. In this social situation, admission to high status groups in direct status hierarchies is not affected by group membership in indirect status hierarchies (e.g. sex, ethnic groups, religious background). In these circumstances, comparisons between indirect status groups would not be socially salient.

The above sketch of the evolution of intergroup relations may sound somewhat optimistic if one compares the final stage with the sort of problematic intergroup relations one finds in most contemporary societies. On the one hand, it indeed seems that the somewhat idealistic perspective is not likely to be fully realized in any society. On the other hand, significant changes have occurred in the past centuries, even in the past decades. Some societies seem already to have achieved a fairly balanced state. It may, however, still take many decades or centuries before a state of dynamic equilibrium constitutes the general rule rather than the local exception.

NOTES

[1] The assumption that status differences in indirect status hierarchies are mainly perpetuated through differential admission to groups in direct status hierarchies seems to be the working hypothesis underlying positive discrimination programmes. Our approach, however, seems to ignore the fact that discrimination also occurs in areas of life which are not related to work. This problem may be overcome by broadening the concept of direct status hierarchies to include social categorizations in terms of e.g. residence areas and schools.

[2] More specifically, the term 'permeable group boundaries', as it is used in the present text, refers to undifferential access of individuals to direct status groups, given that one qualifies in terms of relevant and functional criteria (e.g. ability) and that irrelevant and dysfunctional criteria (e.g. gender, ethnic group) are not used to constrain access to such groups. The impermeability of group boundaries of indirect status groups, e.g. the fact that changing skin color is impossible, is as such not a topic of interest in this chapter.

# 5

# ETHNIC ATTITUDES AND EMOTIONS

Anton J. Dijker*
University of Amsterdam
The Netherlands

At first glance, intergroup perceptions and emotional reactions seem to be related in such an obvious way that the emotional aspects do not seem to deserve any special attention from investigators in the field of intergroup relations. As long as emotions can be seen as logically following from perceptions, people will not see a large difference between the two. For example, it comes as no surprise that the perception of outgroup members as thieves is often accompanied by feelings of fear and distrust, or that a heavy emphasis on an outgroup's culturally deviant conduct may sometimes arouse irritation or even anger in the perceiver. Indeed, modern theories of emotion (e.g., Frijda, 1986; Lazarus, 1984) acknowledge the close relationship between cognition and emotions. People normally react emotionally to events that are appraised in terms of their relevance to well-being. In general, if someting happens that is not in agreement with subjects' concerns, negative emotion will result. Because most of our interpretations of social behavior which confirms or violates social rules are based on shared knowledge (cf. Harré, 1979), most perceived antecedents of emotions are easily verbalized by subjects. Thus, in the case of interethnic relations, subjects are often able to provide 'reasons' for reacting emotionally in the presence of ethnic minorities. From a practical viewpoint, these reasons, to the extent that they are based on false attributions and stereotypes, may be replaced by inferences with non-emotional or even positive emotional consequences.

So far, a cognitive approach to intergroup perception and stereotyping (cf. Stephan, 1985) may not indicate anything special about the emotional consequences of intergroup perception. The correspondence between emotion and perception, however, is not always that simple. Sometimes emotions may be felt for which subjects themselves are not able to provide reasons, even after they have tried hard to find them, or they may feel sometimes that they had 'not enough' reason to react with the intense emotions observed. In these cases psychologists, as well as lay persons, would wonder what aspects of the situation and which personal dispositions of the subject could have produced these reactions. Although one is tempted to say that there is 'nothing cognitive' about these 'unreasonable' emotional reactions (e.g., Zajonc, 1980), we assume that appraisals of a special kind are involved.

In this chapter, it will be argued that interethnic relations, especially, may arouse feelings based on appraisals which are difficult to verbalize. Understanding the determinants of these feelings may be crucial to understanding the elusiveness and perseverance of negative ethnic attitudes. Two classes of determinants will be discussed. The first class involves situational aspects which are characteristic of the kind of interethnic relations to which we will limit ourselves. The second class includes factors which are intrapsychic in nature.

Interethnic relations in the Netherlands have two important properties that make them especially prone to the emotional reactions we are interested in. (I should emphasize at the outset that I discuss these relations only from the majority point of view). First, they involve the presence of ethnic minorities with highly visible culturally and/or racially specific characteristics and behaviors which are somehow experienced as 'strange' by the majority. In the Netherlands, the behaviors, clothing, physical appearance, and beliefs of immigrant workers from Turkey and Morocco are easily distinguishable from those of the autochtonous Dutch. One of the major determinants of these differences is Islamic religion. Surinamese, a third major ethnic minority group, are primarily distinguishable from the

Dutch by their black or dark brown skin colour. Apart from strangeness, the interethnic relations we focus on are characterized by instability. No clear boundaries between groups exist, like those of different caste groups in India, which prevent integration from the outset. Instead, the minority groups involved are trying to bridge the initial gaps between them and the majority, both in a spatial and a psychological sense. From the perspective of the autochtonous majority, this may be seen as a continuous threat to personal space. We assume that this applies even when one cannot speak of 'realistic group conflicts' (cf. Tajfel, 1982). The unpredictability of intergroup relations may pose a need for control over the future development of these relations. I will present empirical evidence showing that situations in which one is confronted with strangers who have a certain degree of unpredictability and uncontrolability may trigger anxiety or irritation. This remains so even when no substantial evidence of bad intentions or norm violations can be found by subjects, or even when those strangers are positively evaluated and subjects are willing to interact with them. Self-reported antecedents of emotional reactions were studied in order to unravel which elements in the intergroup situation probably gave rise to tendencies to maintain distance and aggression. Most of these perceived antecedents or appraisals involved verbalizations of specific reasons for reacting emotionally. In this chapter, however, we will deliberately abstract from the specific content of these appraisals to gain an understanding of the perceived formal properties of the interethnic situation. It is argued that it is the constellation of situational features which may make the antecedents of aroused feelings difficult to understand and which initially arouse feelings of an unreasonable character.

The second class of factors which is probably responsible for the lack of correspondence between emotion and perception is intrapsychic in nature. We will explore the implications of Allport's statement that the "emotion behind prejudice is diffused and overgeneralized, saturating unrelated objects" (1954, p. 204). Socially delicate and unacceptable feelings and actions against particular groups often need to be justified to oneself and others. Therefore, it may sometimes be necesary to transform the original reasons for those feelings and to be in control of emotional expressions. These processes may be called 'regulation of emotion' (cf. Frijda, 1986). I will discuss several processes which can be seen as responsible for the development of stereotypes and the accumulation of 'new facts' about the out-group.

## THE NATURE OF EMOTIONAL REACTIONS TO ETHNIC MINORITIES

In general, research findings indicate that people are quite able to express their ethnic attitude in terms of different emotional qualities. The aim of a first study (Dijker, 1987b) was to identify the relevant categories of emotion. Dutch autochtonous subjects, varying widely in background characteristics took part in a mail survey. They indicated the subjective frequency with which they had experienced different emotions and action tendencies when confronted with members of different ethnic minority groups. They also expressed their general evaluation of the ethnic groups. For both Surinamers and immigrant workers from Turkey and Morocco, four emotional qualities could be distinguished, each with its own characteristic action tendency, and each strongly related to the subject's general evaluation of the ethnic group: 'anxiety' (with a tendency to maintain social distance), 'irritation' (with a tendency to aggress), 'concern' (related to a wish that the outgroup would move from one's neighborhood), and 'positive mood' (with a tendency to approach). A second (Dijker, 1988a) and third study (Dijker, 1988c), which measured the same emotion variables in a student sample and a group of adolescents (mean age 15 years), respectively, largely replicated these findings. In one study (1988a), Dutch subjects reported their emotional reactions to Turkish people [1], in the other (1988c) their reactions to Surinamese. With the exception of 'concern', which did not show up as a separate factor, 'anxiety' (now including the concern items), 'irritation', and 'positive mood' emerged again as emotional categories [2]. The overall evaluation of the ethnic group could well be expressed as a weighted sum of the three emotional qualities. 'Positive mood' was positively related to this evaluation or attitude, 'anxiety' and 'irritation' negatively.

Two further studies (Dijker, 1987a, 1988b) were conducted to investigate how subjects would differentially process information about an ingroup and an outgroup member. Dutch children listened to tape-recorded interviews in which children gave various comments on behaviors of a hypothetical Dutch or Turkish classmate. It was the subject's task to imagine that the target, about whom the speakers on the tape recording supplied information, would become their classmate and to form an impression of him. Although different independent variables were manipulated in the two studies, both showed a main effect for target. Subjects expected to feel more anxiety when confronted with a Turkish target than when confronted with a Dutch target. At the same time, they expressed more positive thoughts about the Turkish than about the Dutch target [3].

Now that we have identified two negative emotional states of interethnic relations – anxiety and irritation-we can be more specific about their nature and the circumstances under which they are aroused. Park (1928) and recently Stephan and Stephan (1985) see anxiety as a basic emotional component of interethnic encounters. According to these authors, subjects feel uncertain whether scripted acitivities and interaction rules and rituals, which normally provide a baseline feeling of control in everyday ingroup interactions, apply in these situations. Barendregt and Frijda (1982) describe anxiety as "the failing of cognition. It is an emotional response upon loss or lack of cognitive grip, cognitive orientation, or cognitive mastery. Correlatively, anxiety is the response upon the impossibility of envisaging appropriate action" (1982, p. 18). In contrast, fear is a reaction to a specific danger to which a specific response of protective behavior is possible. That in the case of anxiety no answer is possible – i.e., one does not know how to handle the situation – is expressed in tenseness and nervousness, its corresponding activation mode. There is readiness to answer the situation, but one just does not know which activities are appropriate to do so. The result is 'blocked acitivity' (Frijda, 1986). Indeed, when subjects interact with stigmatized others (Kleck, 1968), or members of different racial groups (Ickes, 1984; Word, Zanna, & Cooper, 1974), they show the expressive signs of anxiety. Studies in which subjects were free to avoid contact, however, demonstrated a tendency to keep or take distance (Crosby, Bromley, & Saxe, 1980).

An important distinction between anxiety and irritation is the fact that the experience of irritation involves a felt aggressive tendency directed towards the object of emotion, whereas anxiety involves a desire to move away from the object. Irritation may occur when it is clear that avoidance is not possible and the presence of the target will last for a while. Dijker (1987b) found that irritation increased with increasing proximity of minority members to the respondent's residence. Anxiety was not influenced by proximity, probably because anxiety is typically an emotion that is aroused when subjects anticipate the harmful consequences that contact generally may have. Irritation may also occur retrospectively, when one has regained control over one's behavior after an interethnic encounter, with the target's emotion-arousing properties still in mind, or it may even occur during interethnic contact when the expression of aggression is 'safe' for the subject – e.g., when the target of aggression cannot retaliate (Donnerstein & Donnerstein, 1976; Rogers & Prentice-Dunn, 1981). In sum, although anxiety and irritation are different in nature, they may both occur under different cirmcumstances as a reaction to the stressful features of interethnic contact.

*Are ethnic attitudes 'unreasonable'?*

An ethnic attitude may be called a 'feeling disposition', a disposition to experience particular emotions when confronted with the object of the attitude. An attitude can be verbally expressed as an overall evaluation of the relevant social category. The evaluation should correlate highly with emotional reactions and felt action tendencies; we have noted that in fact it does. An attitude may also be conceptualized as a hierarchical cognitive structure which contains the aggregated perceived antecedents of emotional reactions. On the highest level, directly linked to the category label, we would find the most general appraisals. Categorization of a target may then trigger the most global appraisals at the top of the structure. For example, the target may be appraised as weird and frightening, annoying, or enjoyable even before its specific characteristics are interpreted (Dijker, 1988a; Fiske, 1982).

Essentially, this is the kind of non-correspondence between feeling and thinking which I see as responsible for the elusiveness and perseverance of ethnic attitudes. It must be a rather strange experience when outgroup members, about whom nothing is known except their social category, arouse negative feelings. These feelings may appear to be intrinsically related to the target's category, giving rise to the experience of antipathy. When negative feelings are almost automatically aroused upon perception of the target, the underlying appraisals are relatively uninfluenced by supplying different information about the target. For example, the apprehension one can feel for a spider is not undone by knowing that it does not bite, or even that it is afraid of humans. By contrast, the appraisal that a man with a knife will be dangerous to me can be more easily changed by attributing harmless intentions to him, for instance that he is just doing his job as a butcher. Irritation may also be resistant to cognitive modification. For example, it may be aroused by close contact with people who do not act in any identifiable way wrongly, but who draw the subject's disturbed attention by their salient characteristics – e.g., their tone of voice or outward appearance. In those cases, it is difficult to imagine what re-attribution would look like when irritation is not related to behaviors which are specifically directed at the subject. Essentially, the subject observes side-effects of the outgroup's presence, not intended effects. As we will see later, the attribution of bad intentions to the outgroup may be a desirable way out for ingroup members. In the following sections I will describe how feelings which are, in the above sense, unreasonable may originate in the intergroup situation. Here, however, it should be stressed that I do not agree with Zajonc's (1980; 1984) view that, generally, emotion and cognition are essentially different. I like to emphasize that it may be due to a phenomenal quality of particular emotional reactions that feeling is experienced as independent from thinking. Cognitive appraisals, however global and vague they may be, are always involved (cf. Lazarus, 1982; 1984). It is necessarry to identify aspects of the intergroup situation in order to understand the relevant emotional qualities.

SITUATIONAL DETERMINANTS OF EMOTIONAL REACTIONS TO ETHNIC MINORITIES

I shall discuss two empirical approaches to studying the formal aspects of interethnic situations responsible for the triggering of negative emotions. Through the study of adults' self-reported antecedents of emotional reactions to ethnic minorities, we hope to gain insight into the central and more universal perceived aspects of a problematic intergroup relation. Before doing so, however, I will identify important situational factors with the aid of an analogy which at first sight may seem somewhat odd, but whose relevance was also hinted at by Allport (1954): The nature of a human infant's reactions to the approach of a stranger. Although, of course, intergroup behavior is much more complex, I believe the etiology of an infant's reactions to strangers can help us to understand affective consequences of interethnic contact. Here, the formal properties of these situations can be studied in a relatively 'pure' form. The infant's reactions are not confounded by the higher cognitive processes of adults, which are heavily influenced by verbal ability. Its emotional experiences, therefore, are almost immediately contingent upon structural changes in its environment.

*Wariness of strangers*

Anxiety as a distinctive emotional reaction seems similar to the 'wariness of strangers' which can be observed during the human infant's first year of life. Developmental psychologists often make a distinction between wariness and fear (e.g., Bronson & Pankey, 1977; Sroufe, 1977, 1979). Sroufe considers wariness a negative reaction to the unfamiliar or unknown, "due in part to the lack of response options, to not knowing how to behave, or to a loss of control." (1979, p. 486). Fear involves a negative categorization of an event: the event is assimilated to a negative scheme. The phenomenon of infants' wariness of strangers can be considered a useful affective analogue to adults' reactions to strange outgroup members. In general, a stranger can elicit exploratory and affiliative tendencies as well as wary reactions. A stranger is more likely to elicit the former when the mother is present, given adequate familiarization time, and an increase in the infant's response options and possibilities for control. The approach of the stranger should be delayed or gradual and mediated by toys or play. In addition, the infant has to have the opportunity to crawl away; physically restrained babies show more signs of wariness. Whereas a gradual approach increases assimilability, a stranger intruding into the infant's space by reaching for it or picking it up, typically causes negative reactions (see Sroufe, 1979).

For several reasons, these studies are of interest to intergroup researchers. First, they suggest that there are more, or less, advantageous subjective conditions for acceptance of a strange ethnic group in one's evironment. Perceived control over the outgroup's approach behavior and future place in society may be a major determinant of emotional reactions to its presence. Second, the infant studies suggest that the unfamiliar features of an ethnic outgroup alone, in combination with particular social policies of integration, may be responsible for the arousal of anxiety reactions in the autochtonous majority. When outgroup members are seen to approach too quickly, a lack of control may be felt. The approach of outgroup members may be seen as obtrusive or dominating. As we will see below, it is precisely this aspect of the outgroup that is often reflected in the content of ethnic stereotypes. The third, and probably most important, hypothesis suggested by the infant studies, however, is that the interaction between subjective conditions and characteristics of the stranger may prevent negative affect. In a sense, both parties must make themselves ready to be in each other's presence. For example, an objectively gradual influx of an ethnic minority in a quarter may still arouse anxiety and irritation when the native inhabitants do not feel competent to interact with its members. Conversely, a rapid approach may undermine an already acquired feeling of competence.

*Perceived antecedents of emotional reactions*

Which perceived aspects of the interethnic situation make adults feel anxious, irritated, or happy? An earlier-mentioned study (Dijker, 1988a) measured not only emotional qualities, but also the perceived antecedents or appraisals of these emotions. In the first part of this study, subjects were asked to report freely on the antecedents of experienced emotional reactions to ethnic minorities. The second part aimed at a more quantitative assessment of the structure of outgroup appraisals. A large number of rating scale items were constructed to represent the main appraisal categories found in the first part. A new sample was asked to indicate how often these appraisals had aroused emotional reactions. Factor analysis identified ten general appraisal dimensions, of which the major negative ones were 'perceived egocentrism', 'perceived ethnocentrism', and 'general concerns' (the latter dimension appeared in the present study as an appraisal dimension). It is hypothesized that, at an abstract level, negative appraisals of an ethnic minority reflect subjects' experience of lack of control and predictability in the interethnic situation. Egocentrism might reflect the perception that strangers are unwilling to be influenced by members of the ingroup. In the eyes of the subject, the strangers show themselves as too independent and self-centered. They are perceived as loud, arrogant, obtrusive and untrustworthy. In the interpersonal domain in general, the reasons most often mentioned for disliking other persons are conceited and arrogant behavior – clearly signs that they place themselves outside the social order (cf. Kidd, 1958). The appraisal of ethnocentrism is just an extrapolation to an explicit intergroup level. One sees the outgroup as a closed, impermeable system with unknown intentions and unresponsive to adaptation demands. Finally, general concerns about future developments of the intergroup situation express the unpredictability and anxiety-arousing properties of these developments.

Although it is a large step from infants' reactions to strangers to adult behavior, I believe that adults' outgroup appraisals express, in a more elaborate way, the same kind of affective experience infants have when confronted with a stranger. In both cases, unstable and unpredictable interpersonal relations are involved.

*Do ethnic stereotypes have universal properties?*

How is the content of adults' ethnic stereotypes related to general aspects of interethnic relations? A theory of stereotype content could answer this question. The aim of such a theory is to delineate the universal determinants of the intergroup situation that give rise to basic intergroup appraisals. Tajfel (1982) observed: "Purely cognitive studies need to be supplemented by a theory of the contents of stereotypes, particularly as we know from historical and antropological evidence (...) that the diversity of patterns or types of intergroup stereotypes is fairly limited." [Tajfel's italics] (p. 22). Bettelheim and Janowitz (1950) proposed such a theory from a psychoanalytic viewpoint. I believe that the analysis of emotional appraisals may reveal the subject's perception of the total (intergroup) situation, including both characteristics of the target of the emotion and the subject's own possibilities for action and control in a given situation. Several universal aspects of these perceptions, which correspond with my own findings, are suggested by Campbell (1967).

Campbell (1967) remarks: "If most or all groups are in fact ethnocentric, then it becomes an 'accurate' stereotype to accuse an outgroup of some aspects of ethnocentrism. This generates a set of 'universal' stereotypes, of which each ingroup might accuse each outgroup, or some outgroup, or the average outgroup." (p. 823). He gives some examples of correspondence of self-descriptions and outgroup stereotypes about aspects of ethnocentrism (Campbell includes egocentric and ethnocentric motives within the same category): "We revere the traditions of our ancestors" versus "they love themselves more than they love us"; "we are loyal" versus "they exclude others"; "we are brave and progressive. We stand up for our own rights, defend what is ours, and can't be pushed around or bullied" versus "they are aggressive and expansionistic. They want to get ahead at our expense". It is imaginable, then, that interethnic relations may generate stereotypes in which ethnocentrism is a basic element. The more egocentric and ethnocentric motives one ascribes to an ethnic minority, the more lack of control subjects seem to experience. Like the infant's distress reactions to strangers, adults' demands for integration and adaptation of minorities to Dutch ways of living can be seen as expressing a need for control. This need may generate the basic appraisal dimensions of the intergroup situation. If it is true that this situation has formal properties comparable to the infant's encounter with strangers, it is likely that feelings are produced that, even for adults, may initially lack clear reasons. As the often rich content of stereotypes suggests, however, the appraisals underlying these feelings are easily filled in by concrete behavioral evidence – to which process we now will turn.

## INTRAPSYCHIC PROCESSES

Allport remarked: "The emotion behind prejudice is diffused and overgeneralized, saturating unrelated objects" (1954, p.430). The cognitive content of the stereotype adapts itself to the prevailing feeling tone and the demands of the situation. Any justification of one's dislike that fits in a communication about the outgroup, will do (ibid., p. 204). Like Allport, Park (1928) also saw the stereotype as a rationalization. With respect to interracial contact, Park writes:

"We do not know what, under certain circumstances, a creature so unlike ourselves will do. Even after a prolonged and rather intimate acquaintance with an individual of another race, there usually remains a residue of uncertainty and vague apprehension, particularly if the stranger maintains a reserve that we cannot fully penetrate. Under such circumstances it is inevitable that rumors and legends will arise and gain general currency which purport to describe and explain racial differences, but in fact serve merely to give support to apprehensions and vague terrors for which there is no real ground in fact. Anything that tends to make a mystery of divergent and alien races, even biological theories which suggest remote and ill-defined dangers of contact and intimacy, tends to intensify antipathies and lend support to racial prejudices". (p. 238-239).

People who experience emotions with global appraisals, like those of moods, are likely to construct an appraisal of their environment which is a complement of the experienced action tendency or activation mode of their feeling state. Consequently, people who are in an anxious mood are more ready to perceive various kinds of dangers. This was nicely demonstrated in an experiment by Murray (1933), who called this phenomenon 'complementary projection'. Children who were scared appeared to judge photographs of strangers as more malicious than children who were not scared. Recent research demonstrates that moods have the property of 'saturating unrelated objects' because they are 'diffused', as Allport noted (cf. Bower, 1981). For example, such unrelated issues as nuclear plants and the presence of ethnic minorities in society may simultaneously be seen as the causes of an anxious mood. Murray considers attitude (or 'sentiment') to be a dynamic concept. One of the psychological processes involved he calls 'aggregation': "The assembling in the mind of many images with similar affective meaning" (quoted in French, 1946, p. 258). The result of aggregation is "a loose array of seemingly valid facts, assumptions, and arguments" (ibid., p. 257). The only stable feature of the attitude, then, is a felt antipathy to the object, for which the reasons are interchangable and not of much importance.

Whereas situational factors may produce feelings whose origin the subject has probably never been able to understand, intrapsychic processes may produce a pile of accumulated 'facts' which subjects are no longer able to see through. Both situational factors and subsequent information processing work in combination and produce emotions that are relatively independent of cognitive control. Thus, they are not necessarily meaningfully attached to observed behaviors of the target, but instead may have an 'unreasonable' character. It should be noted that these older psychodynamic conceptions fit well with a modern emotion-theoretical framework (e.g., Frijda, 1986; Lazarus, 1984). Emotions can be considered 'emergency reactions'. Being signals of events that have consequences for subjects' well-being, they activate people to take appropriate actions. Although emotions are in this sense interrupting rule-governed social interaction, they almost never occur without some form of control or regulation. Just 'letting an emotion go', without observing what undesirable consequences this could have, would be quite maladaptive behavior. People can have different reasons and different ways to regulate the emotion process. Regulation can be motivated by the subject's positive or negative evaluation of an emotion. Fear, for example, can be very undesirable for most people because its experience may point to personal weakness. One can try to inhibit the impulse to flee, or postpone the expression of fear, until one is out of other peoples' sight.

These two aspects of Frijda's (1986) emotion theory, the subject's evaluation and regulation of emotion, offer a special entry into understanding the functioning of attitudes. Especially in the field of interethnic relations, where the consequences of negative attitude, like discrimination and hostility, are explicitly and publicly judged, we should expect the subject's evaluation and regulation of the emotions associated with ethnic attitude to play an important role. Negative evaluation of negative feelings against an outgroup can lead to shame or fear of expression. Positive evaluation of negative feelings can lead to acceptance and pride. These meanings, which are attached to the emotions themselves, can have demonstrable consequences for the way in which one regulates one's emotions and attitude. Negative evaluation, which is to be taken as the normal case in our society, could lead to regulation by which one tries to suppress or hide the negative emotions. This does not mean, however, that these feelings cease to exist or that regulation attempts are succesful. For example, efforts at self-regulation may cause observable tense and unfriendly behavior (Word, Zanna, & Cooper, 1974; Ickes, 1984), or one may seek to express negative feelings in indirect ways, or in an extreme, amplified way when a normal sanction is justified (compare the accounts of so-called "reversed discrimination" (Gaertner, 1976), and "ambivalence" in interracial contact (Katz, 1979)). Explicitly positive evaluation of feelings of antipathy, aversion and hatred is found in fascism. Here, negative feelings against ethnic minorities are glorified, encouraged and spread. In fascism and racism, the 'genuineness' and 'undoubtedness' of these feelings have an important meaning (cf. Sartre, 1946). Especially in times of distressing events, such as unemployment or war, one can be certain about who is 'the enemy'. A society is justified in disapproving of the expression of negative ethnic attitudes. It is important, however, to find out how people can regulate the associated emotions, which are, of course, not so easily controlled by them as their expressions.

Regulation of emotion can also be seen as an attempt to ascertain what kind of persons the outgroup consists of, in order to cope efficiently with them cognitively and behaviorally. Reappraisal, or the transformation of the original appraisal component of an emotion, can work as a palliative in case of unpleasant emotions (cf. Frijda, 1986; Lazarus, 1966). Rationalization, for example, which Allport (1954) held to be one of the main functions of stereotyping, can be seen as a more 'advantageous' (re)appraisal than the primary appraisal of the emotion-eliciting event or object. In the context of interethnic relations, this would mean an attempt to attribute specific negative meanings to a strange and unfamiliar outgroup in order to explain feelings of uncertainty and anxiety. While 'intellectualization' (Lazarus, 1966) works against the unfolding of unpleasant emotions (the subject maintains distance by adopting an objective view of the situation), the reverse may be true of certain kinds of projection. Because of the attribution of responsibility and, on an intergroup level, of bad intentions to an outgroup, the emotion of anger becomes possible. Interestingly, there are indications that projection of intentions can work in a stress-reducing way (Sherwood, 1981), presumably because the resulting anger implies the experience of control (Novaco, 1979). Thus, when continuously annoyed by the appearance and characteristic behaviors of a certain ethnic group, anger based on accusation is a more advantageous emotional state. At least it reassures people that they are responding to things that are under their control.

## PRACTICAL USEFULNESS

Tajfel (1969) advocated the view that, in contrast to a cognitive approach, a consideration of prejudice as an emotional phenomenon was "useless in the planning of any form of relevant social change" (p. 190). In light of modern theories of emotion, however, a more optimistic voice may be raised. First of all, emotions are reactions to events that are important to the subject's well-being. Emotional experience should, therefore, be taken seriously in order to come to personally relevant perceptions. In this way, we may discover perceived aspects of the intergroup situation which are the actual basis for maintaining social distance or becoming hostile. These perceptions are the ones we have to influence if we want to fight harmful images of ethnic groups. But, as is apparent from our discussion, this may be a difficult task for several reasons. To the extent that those images are not based on specific aspects of an ethnic group, but are more contingent on a whole configuration of situational features, it is the perception of whole situations that must be altered. For example, one may think of heightening the experience of control over the intergroup situation by improving subjects' coping abilities and self-efficacy during real or simulated interactions. Or, more symbolically, one may try to reduce feelings of uncertainty and concern about future developments of the intergroup situation by supplying information about those developments. Recall, too, the importance of an adequate familiarization time. Awareness of situational determinants of negative interethnic emotions should lead to the selection of those integration strategies that produce the least negative emotional side-effects.

Elsewhere (Dijker, 1987b), I also argued for the "disconfirmation of negative sentiment". That is, instead of invalidating subjects' stereotypes by means of argument, one should acknowledge that negative attitudes – being dispostions to experience negative emotions – probably have their own validation strategy. Ultimately, their subjective validity may only be disconfirmed by supplying conditions for the experience of strong positive emotions (Dijker, 1987b). These conditions can be formulated on the basis of subjects' accounts of successful and positive interactions with members of ethnic groups (i.e., reasons for positive emotions).

Apart from an emphasis on the intergroup situation, we should also focus on the intrapsychic processes responsible for the gradual divergence of feeling and thinking about ethnic minorities. Prophylactic use of principles of psychoanalytic theory may be useful, even though a mass psychoanalytic treatment of prejudiced subjects is unrealistic. For instance, one can think of explaining the ways in which emotions can be handled in different situations. The automatic control processes may then come under conscious control. More concretely, one could learn when rationalization or projection are active, reading such material as Allport's (1948) ABC's of Scapegoating. Of course, presentation of these materials should be adapted to the population they are intended for.

Attitude is a dynamic concept. Cognitive approaches to prejudice acknowledge this point, but they typically emphasize subjects' activity and flexibility in the domain of knowledge, while they attempt to retain an easy and coherent picture of the outside world. Neglected, however, is an integral account of the regulation of thoughts, bodily changes associated with emotion, and expressive behavior. It may, for example, be important to know what happens in thought when attitude expression is not allowed. As suggested earlier, knowing the relevant regulating mechanisms seems to be especially important if one wants to address the public in a non-harmful way about the issue of ethnic prejudice, discrimination, or racism. At least one should make a clear distinction between hostile expressions of ethnic attitude and the underlying negative feeling. The first we rightly forbid, but by not distinguishing it from the latter, one disapproves of or even denies a part of the subject's reality that can be very painful indeed - the *existence* of negative feelings toward minorities. It is undesirable to await the expression of these slumbering feelings until large-scale distressing events mobilize them and render them largely uncontrollable. An important question, therefore, is what kinds of regulation constitute efficient coping with negative feelings against ethnic groups.

In this chapter, the importance of emotional aspects of ethnic attitudes was stressed. Admittedly, no research has been done which systematically tries to evaluate the practical usefulness of the presented view against existing practices to influence attitudes. I hope this line of investigation will be followed in the future.

NOTES

* The Netherlands Organisation for Scientific Research is gratefully acknowledged for funding this project. This research was performed with the support of a PSYCHON grant from this organization (560-270-006). I would like to thank Tom Pettigrew for his helpful comments on an earlier draft of this chapter.

[1] This study also measured the perceived antecedents of emotional reactions to minorities. The results will be discussed in a later section.

[2] A possible reason for this discrepancy may be the different nature of the concerns of the samples used. The general population of Amsterdam may be more involved with issues that typically arouse the emotion of concern – e.g., unem-

ployment seen as caused by the presence of ethnic minorities and harmful future developments of interethnic relations. By contrast, the student sample and the sample of adolescents may react more emotionally to concrete harmful encounters which cause anxiety or irritation rather than concern. Although this sample may be negatively affected by issues such as unemployment, these may not be salient enough to arouse the distinguished emotional state of concern. This chapter will concentrate only on anxiety and irritation.

[3] This main effect, however, was qualified in both studies by interactions with other independent variables. In one study (Dijker, 1987a), ambivalence was only shown when subjects had to process information under high time pressure and were held accountable for their final judgments at the same time. In other words, ambivalence emerged in those conditions in which information processing was difficult for subjects. Another study (Dijker, 1988b), which manipulated mood state as a second independent variable, revealed that there was no differential anxiety toward the ingroup and outgroup target when subjects were in an angry mood. This finding was explained by assuming that the emotion of anger has a characteristic aggressive action tendency which works counter to the experience of anxiety.

# Part II

# Real Life Studies

# 6
# ETHNIC IDENTITY
# AND THE PARADOX OF EQUALITY

Sawitri Saharso
University of Amsterdam
The Netherlands

Much has been written about the development of identity among ethnic youth [1], though empirical research on the topic is relatively sparse. Where research is carried out, little attention has been paid to the social context within which the identity development takes place. The study reported here is aimed at attaining a greater understanding of the link between ethnic identity and its social context [2].

In the following pages, I shall discuss the responses of ethnic youths when confronted with the question, 'Who am I?'. I shall also discuss the various ways their origins may influence the form they give to their lives, as well as their views on their friendships, their choice of future partners, and the perception of their life-chances. Finally, I shall consider the reactions of the (Dutch) environment, the school environment in particular, to the youths' ethnic origins and the link between that reaction and their own self-perception.

## THE CONCEPT OF ETHNIC IDENTITY

Various notions have been developed concerning the concept of 'ethnic identity'. In the literature, a distinction is often made between 'individual identity' and 'social identity'. This distinction is generally related to differences in academic disciplines; e.g. a psychologist is more concerned with *individual* processes of self-perception, the anthropologist more with the *social* aspects of group formation and interaction (see e.g Epstein, 1978, pp. xi-xii). The concepts of individual identity and social identity, however, are not mutually exclusive. Groups are clearly composed of individuals, and group formation can only take place through the personal identification of individuals with the group. In this study, I shall focus on the social aspects of ethnic identity, using Tajfel's definition of social identity as:

"that *part* of an individual's self-concept which derives from his knowledge of his membership of a social group (or groups) together with the value and emotional significance attached to that membership." (Tajfel, 1981, p. 255).

This 'social' conception of ethnic identity entails that 'ethnicity' is not necessarily a central factor in the way individuals experience their identity. It also avoids presumptions concerning the way in which individuals (eventually) give meaning to their ethnicity. I do assume that ethnic identity cannot be studied isolated from social context. The importance of interaction with the social environment for the formation of ethnic identity was aptly illustrated by Tajfel (1978, pp. 6-7) who described the experience of West Indian students in England: Only after they had realized that, irrespective of who and what they were as individuals, they were, in the eyes of British society, in the first place 'blacks' – where 'black' stands for low status and the corresponding treatment by society – did they develop a 'black consciousness'. Both the content of ethnic identity and its significance in relation to other aspects of identity (such as gender, class, and generation), are to a large extent dependent upon the reactions of the environment (see also Brake, 1985, and Weinreich, 1979).

For the present study, 'ethnic identity' has been operationalized in four ways. Ethnic youths were asked to consider the role of ethnicity in their self-perception, in their choice of friends and their (future) partner, and in the perception of their individual life-chances. The study was conducted in two Dutch secondary schools. The 'reactions of the social environment' were comprised of the reactions of fellow students, under the assumption that at school these reactions largely constitute the social context for ethnic youths. In other words, for ethnic youths, the way in which native Dutch youths react to them can be important in determining how they experience their ethnic identity. The ethnic identity of native Dutch youths will not be dealt with in this chapter, as I shall focus on the identity of adolescents with other ethnic origins.

## ETHNIC GROUPS IN THE NETHERLANDS

The immigrant population of the Netherlands is composed of a wide variety of groups. As has been explained in the introductory chapter, a distinction can be made between migrants from the former Dutch colonies and migrant laborers – the latter being mainly of Turkish and Moroccan origin. While migrant youths from both categories participated in the study, the cases selected for discussion in this chapter all find their origins, with one exception, in the former colonies of Dutch East India (present day Indonesia including the island group of the Moluccas) and Dutch Guiana (Surinam).

Let me briefly sketch the history of migration from these former colonies. During the period of decolonization, many people from the (former) colonies moved to the Netherlands. Indonesia gained independence in 1949 and in the course of the following ten years approximately 195,000 Moluccans, Chinese, Eurasians (Indo-Europeans), and a small number of migrants of solely Indonesian origin (from which group I originate) moved to the Netherlands. Surinam gained independence in 1975, with the major portion of the approximate 250,000 migrants of Surinamese origin coming to the Netherlands in the late 1960's and 1970's. The Surinamese migrants also have diverse ethnic origins: Afro-Caribbean (Creole), Hindustan, Chinese and Javanese. The ethnic composition of the Surinam population is a direct consequence of Dutch colonial policies. First, Africans were 'imported' as slaves to work the plantations; the Creoles are their descendants. The Javanese were recruited as contract labor from the Indonesian island of Java in the nineteenth century, followed by Hindustani recruits from the Indian sub-continent. The groups display considerable differences, especially in religion and language. Most of these colonial migrants do possess the Dutch nationality and speak Dutch to varying degrees of proficiency. Another characteristic they hold in common is that they clearly differ in physical features, such as skin color, from the 'native' Dutch.

## DESIGN OF THE STUDY

The analysis is based on the methodological principles of 'grounded theory' as developed by Glaser and Strauss (1967). This is an interpretative approach to the conceptualization of how people perceive reality, in which theory is systematically generated from (grounded) empirical data by means of comparative analysis. Procedures are obviously required to check the reliability and validity of the analysis results. For our purposes, method and data triangulation procedures were applied: Results derived from various methods (of data accumulation and analysis) were checked for consistency, as were the results arrived at for different cases (i.e. schools). A check was also conducted on deviant cases; the exception must be explained as well as the rule. Strictly speaking, I can only draw conclusions for the specific schools observed, though I shall attempt to reinforce the credibility of the results by indicating where they correspond with the insights gained from other research.

This chapter draws upon research carried out at two schools for secondary education geared to vocational training. The schools were selected because of the heterogeneous ethnic composition of the student population (in our broader research framework, a total of five schools have been observed to date). During a two-month period, a senior class at each school was observed during social studies lessons, in which ethnic relations functioned as theme through lessons on prejudice and discrimination. Besides these observations, the results are based on document analysis and (in)formal interviews. The documents were compositions written by the students in which they present themselves to a (fictional) pen pal. Sixteen students, eight from each class, participated in structured interviews, which were recorded and later transcribed. My own ethnic background was probably significant inasmuch as the students with ethnic origins in Indonesia, and the Javanese and Hindustani students from Surinam regarded me as a member of their in-group. My typically Javanese surname and the fact that my first name occurs fairly frequently among Hindustanis played a role, as well as my appearance and behavior. My impression is that this led to a greater candor on the part of those ethnic students interviewed.

One class was composed of thirteen students, including a Javan/Surinamese boy, a girl with a Moluccan father and Indo-European mother, a Hindustan/Surinamese girl and a girl with an Indo-European father and Dutch mother. The second class consisted of eighteen students, including a boy with an Indonesian mother and a Dutch father, a girl with a Dutch father and a Spanish mother, and a girl with a Creole father and a Javan/Indonesian mother. The remainder of the students were Dutch, though one girl had a Yugoslavian step-father. All students were born in the Netherlands or had immigrated before their fourth birthday; at the time of the observations their ages ranged between 17 and 19. In the results presented in the forthcoming section, the students, for the sake of authenticity, are quoted as literally as possible. Some clarification of ethnic terminology may then be called for: Moluccans are also referred to as 'Ambonese', after the main island of the Moluccas, while Indo-Europeans are referred to as 'Indos' or 'Indies'. The Creoles are referred to as 'Negroes', or, being the ethnic majority in Surinam, as Surinamese. Whenever race relations are referred to, the students, regardless of their ethnic origin, do not use the terms 'ethnic' or 'migrant', but speak instead of 'foreigners' and their 'Dutch' counterpart.


RESULTS: THE ETHNIC STUDENTS

*Self image*

The first question put to the seven students in the interview was: "Tell me something about yourself; who are you?" They generally reacted by stating how old they were, what their hobbies were, and how many brothers and sisters they had. Practically no one said anything which referred to their ethnic origin, one boy illustrating the exception:
"I live in Purmerend (a town 20 km from Amsterdam, SS) and I've lived here for 15 years. I came here when I was two. To the Netherlands."
The compositions they wrote about themselves provided similar information. This would suggest that ethnicity plays only a subordinate role in their self-perception.

If the students said nothing about their origins in response to the first question – as was usually the case – I then asked them directly about their origins. It was apparently an unpleasant subject to dis-

cuss. Here is an illustration from an interview with a girl, Dominique:
"My father came from Ambon and my mother from Ceribon" (city on the island Java, SS).
Question (Q): "As your father came from Ambon, is he Moluccan?"
Answer (A): "Yes."
Q: "And your mother too?"
A: "No, she's a real Indo."

When I asked whether other people at times ask where she comes from, she answered:
"They see I'm Indies."
I returned to my question:
"What do you say when people ask what you are?"
Now she answered:
"I don't really go into it much. They just ask something and then I say it and then they don't really delve into it much either."
When I pressed her further, she answered in the end:
"My parents come from Indonesia and I was born here. In Zaandam" (another town not far from Amsterdam, SS).
So Dominique sometimes describes her origin as "Indies", sometimes as "my parents come from Indonesia", and when talking to me, an 'insider', she describes her origin in terms of the places her parents come from. Her answers seem to depend on to whom she is speaking. We can, however, derive more information from Dominique's answers. She can describe her ethnic origin in various ways, but she does not describe it as 'Moluccan' or 'Indonesian', and she immediately adds, once her parents' Indonesian origin is noted, that she herself was born in the Netherlands. It is also clear from her answers that Dominique is used to warding off questions about her ethnic origin. She evades them, her answers being in effect a warning not to go on asking.

I met with the same reaction from the other students as soon as I pressed my questions; they were ill at ease, wary, and even evasive. This reaction would appear contradictory, were their ethnic identity really unimportant to them. Why the need to keep it so protected, or rather, to keep it hidden from view? Besides the evasive behavior, there are other indications that ethnicity plays a role in these youths' self-perception.

*The situation in the class*

The respondents were asked how they liked being in a class of mixed ethnic origin. Their reaction was generally one of indifference, indicating that it did not matter one way or the other. The ethnic composition of the class gained in significance when the question was posed in a slightly different manner: "How would you like to be the only 'foreigner' in the class?" All the respondents were clearly opposed to such a situation because they would then feel like an exception. Grace, the girl with a Creole father and Javanese mother, felt that she was already an 'oddity', even with Peter, the 'Indies' boy, and Irene, the half-Spanish girl, as classmates. Grace did not feel comfortable in her class.

The observation data indicated that the seating pattern in the classes is ordered along ethnic lines. Ray, for example, of Javan/Surinamese origin, usually sits next to Rinnia, the girl of Hindustani origin, despite the fact that boys normally sit next to other boys, and also despite the fact that Ray otherwise avoids contact with Hindustanis. Ray on this subject:
"Look, Rinnia is Hindustani, but I've known her since the first grade; I know how she is. I just go around with her, the way she is. But, normally, I don't go around with Hindustanis."
Rinnia's best friend is Yvonne, a Dutch girl. These two form a clique with the remaining two students in Class 1 with a migrant origin – the 'Indies' Dominique, and Nannie, whose father is Indo-European. In Class 2 we observed that Peter, the 'Indies' boy, always keeps company with Irene, the 'half-Spanish' girl, while during the breaks Grace seeks the company of a Surinamese girl from another class instead of the girls of her own class.

We see that these ethnic adolescents tend to draw together in school classes where they constitute a minority (less than 25%) of the students. This attraction is apparently motivated by the social situation in which they find themselves as a minority: where they refrain from contact in other circumstances they associate with each other here, even if this implies overlooking the prevailing gender barriers in the class. The choice of social contacts is then determined more by the common factor of migrant origin than by the common factor of gender.

*Friendship*

Ethnicity also appears to be important with regard to friendships. An interesting biographical aspect conveyed by several of the ethnic youths during the interviews was a clearly defined moment of choice, where they were confronted with the issue of the ethnic nature of their friendships: 'Do you want to belong to us or not?' Peter (Indies) told this story:

"I was, I think, eight or so and some white kids and Indies kids were having a match, building a dam to the other side, who'll be the first on the other side. I was with the white boys, as I always went around with them. But this was against Indies boys, a lot of Indies and some others, Turks. (They called) 'Hey, come here, you're a colored too, aren't you.' I said, 'Why? I always go around with them and now I should help you guys. It makes no difference. I'm staying here, so there. That's the way I am.'"

Even when appealed to on the basis of his ethnic origin, Peter chose his white Dutch friends. When people ask him what he actually is, he sometimes responds with, "I'm Dutch", referring to his nationality. Peter continued:

"But usually they ask whether I'm Indies, or a half-blood, and then I just say, 'yes, that's right'."

Some of the students have only ethnic friends. For some of them, the preference – who they want as friends and who not – is directly related to ethnic factors. Here's Ray (the Javan/Surinamese boy):

"I mostly go around with my own type. Surinamese and Javanese, Indies people. I don't know why; that's just the way it is."

Q: "Do you feel more at home with them?"

A: "Yes."

Q: "But not with Chinese from Surinam?"

A: "No, I've never been friends with a Chinese boy."

Q: "And Hindustanis?"

A: "No, not them either, not at all (...) I don't go around much with Hindustanis; I don't know why not, I just don't".

Ray's contact with Surinamers or Negroes (he uses both terms) leads to conflicts with his parents. Ray:

"I just go around with them, I always have. My father and mother, they don't really like Negroes, but I do. I think you should simply go around with them, but my father and mother don't want it. Well, they do, but I cannot bring them home."

"Look, I can take them home, the Surinamers, the guys, but then my

mother just gives the cold shoulder you know. (...) She doesn't like it."
As far as the Dutch are concerned, Ray says that he does not make friends with them very easily. Not that his mother would object, although she prefers that he take Indies or Javanese boys home with him. She really only objects to his contacts with Creole (Surinamer) boys.

Ethnicity clearly plays a role in Ray's choice of friends. In view of the above, one can hardly maintain the assumption that he is unaware of this fact. Otherwise he would not be able to talk about his choice as he does. However, he still claims that ethnic origin does not interest him. Ray does not experience what he says about (ethnic) friendships as contradictory.

*Partner choice*

When asked what kind of person their future partner might be, the students often used ethnic categories. Dominique responds to my initial query by conveying that her future partner could be anyone. I then suggest:
"A Turk or Surinamese man?"
Then her 'anyone' is shown to have its limits. Dominique (shocked):
"No, never that. That just isn't my taste. I never look at such boys. I look more at Indos and Dutch. Honestly, no."

Dominique's parents also have their preferences, which may explain why Dominique does not call herself Moluccan or Ambonese:
"My father says I shouldn't marry an Ambonese, but an Indo or Dutchman. Preferably someone Dutch, but not of their own races; they don't want that. At least my father doesn't want that, my mother doesn't really care."
Rinnia also knows exactly what she does not want: a Hindustani husband. Because, according to her, "they are so different, have other ideas about women, don't allow you very many rights". Rinnia already has a boy friend, "a very sweet boy, half Indies, half Italian", a friend of Ray's. Rinnia:
"He went with a Dutch girl for two years, but he doesn't see much in Dutch girls. They always think they know it better, and they're so bossy. I'm shy and quiet, just like him. Quite different from Dutch girls."

She thinks that's why he fell for her and she hopes the relationship will amount to something.

*Life chances*

These adolescents have all experienced in their daily lives that their ethnic origins influence the way in which they are treated. All of them have had personal experiences with discrimination. From the examples they give, these apparently involve aggression on an inter-personal level: They have been sworn at, threatened, and drawn into fights. How do these students generally view their chances in society? Rinnia on her career opportunities:
"If they have the choice between a Dutch girl and me for a job, then they surely prefer the Dutch girl."
And, speaking literally of her opportunities in life, referring to the murder of an Antillian boy:
"Well of course it could happen to me, it can happen to any foreigner."

The situation for Peter is more complicated. We already saw how in another situation he chose for the 'Dutch' side. One is not supposed to discriminate, and Peter will not readily accuse Dutch people of discrimination. Within the logic of this moral framework, Peter should deny that he is subject to discrimination. He does deny it. When asked whether he thinks he has less opportunity in society:
"I don't think so, really."
But he goes on to say:
"At the most you might have less opportunity because others have messed it up for you, who are brown or have another na(tionality), origin. (...) It's just like a white boy of 17 coming and messing it up, and then another boy of 17 messes it up, and then the following boy of 17 is rejected out of hand. And I think it goes like that with coloreds."
Peter does not want to accuse the Dutch of discrimination, but at the same time he is aware that reality, in this respect, can be disappointing. He makes the one view compatible with the other by applying a reasoning tantamount to 'blaming the victim'. To summarize: Irrespective of the reasons they give, all the interviewees are aware that they have fewer opportunities in society than Dutch youths.

*Conclusion and discussion*

For these adolescents, thinking in terms of ethnic categories when talking of friendship and love relationships is apparently a matter of course. Moreover, they discuss these subjects with their parents in the same terms. Ethnicity seems to be of great significance to them when laying down the patterns of their lives: whom they sit next to in class; whom they associate with; whom they think to share their life with. Furthermore, they are distinctly aware of their subordinate position in society. Here too their self-image and social expectations are colored by their origin. One would expect these youths to perceive their ethnicity as an important aspect of their personality. This would fit with Deschamp's findings, where he states:

"Being placed in a position of minority or of being dominated produces in the individuals involved a heightened awareness of the social categories which determine their minority status" (Deschamps, in Tajfel, 1982, p. 91).

At first glance, however, little could be noted of this 'heightened awareness'. Practically no one referred to their ethnic origin when asked to convey who they were, but all made a remark at some point in the interview which inferred that their appearance or origin was not important, for instance:

"I don't dislike anyone in particular. No race or such. Makes no difference to me. We're all people."

or:

"I just see everyone as people, to me everyone's equal."

Their choice of friends and partners, however, demonstrates that 'race or such' is definitely a relevant classificatory criterion for these adolescents. They give the impression that one ought to consider ethnicity an irrelevant characteristic and that that is how they would personally prefer to view the subject. Apparently there is a standard here which they subscribe to at face value, but which they cannot always uphold in their further views and behavior.

Of course, such a discrepancy between (superficial) views and behavior is not in the least uncommon. Most people prefer to think they arrive at their choices in life autonomously, for instance in the choice of partners. Just as 'romantic' beliefs, according to the myth, overrule more germane social interests by partner choice, these youths probably prefer to think they have chosen their particular friends 'just because they like them'.

However, this does not explain why they think that ethnic origin is a factor which ought to be overlooked. Social standards ideally serve specific purposes, but what is the purpose of the standard reflected by the statements 'we are all humans' or 'everyone is equal'? In my interpretation, these statements refer to the principle of human equality 'Irrespective of appearance or origins, people ought to be treated on the basis of equality'. These ethnic students do not expect that they will be treated according to this standard, and they are reacting to this: 'Discrimination on the basis of race or ethnicity is also wrong; while people like ourselves are discriminated against on these grounds. Ethnic origin makes no difference to me and it should not make any difference to others.' This would prohibit them from saying that ethnicity plays a prominent role in their self-perception and in the social patterns of their everyday lives, as it would be in direct contradiction to the principle of equality they (would like to) believe in. Together with the desire (or need) to perceive themselves as (autonomous) individuals, a barrier is thrown up against (self) perceptions emphasizing ethnic differences.

We must now confront the issue of why these students are on their guard when the subject of ethnic differences is brought up. Why do they have (negative) expectations with respect to society treating them as equals? Have they had such bad experiences in relation to their ethnic origins? What attitudes does their environment usually adopt towards them? As I am unable to obtain an overall view of their social environment, I shall limit the discussion in the context of this study to observations of their fellow students with Dutch (ethnic) origins.

## THE DUTCH STUDENTS

### The Dutch on their ethnic fellow students

Most of the Dutch students displayed a 'neutral', or even indifferent attitude on the subject of 'foreigners'. In response to the question of what they thought about the fact that migrant students were in their class, they mainly shrugged or said it made no difference. Our observations on friendship patterns correspond with this picture; Dutch and migrant students sit alongside each other in the classroom, not with each other. Interethnic friendships such as the one between Rinnia (Hindustani) and Yvonne (Dutch) are rare. Rinnia told us how she and Yvonne became friends, with some remarks in passing on how their environment reacted:

"Everyone says it, 'how's that possible, you two becoming such good friends'. Even Pieter (a teacher, their class counselor, SS) says it. (...) She (Yvonne) thought it fun to have a Hindustani girl friend for once. She thought it would be different you know. Well, nobody said anything, or whatever, but they seem a bit taken aback. Like 'can you get on with her so well' and so on. Because if she tells her friends then she says 'her name's Rinnia', and then they always ask 'what is she?' ... you know ... and then she says 'a Hindustani girl' and then they look surprised. It's very nice that she can get along with me so well. It's nice to hear that."

Q: "And you, were you surprised you could become such good friends with a Dutch girl?"

A: "I've always been able to get along with girls. I've never had any trouble with that. I don't always try to make contact, you know, but I can always get along well with everyone."

There is something odd occuring in this interview fragment. Two sentences are enough for Rinnia to deal with herself and the reactions of her surroundings. Apparently there is nothing exceptional in anyone's eyes, including her own, in her seeing Yvonne as a friend. Two remarks: Rinnia refers vaguely to 'girls', and we hear she can get along with everyone. We have heard this message before. What is new however is the asymmetry in the reactions coming from the surroundings. The reactions do not arise in Rinnia's environment, but in Yvonne's environment directed at Yvonne. Yvonne goes against the current of her environment's expectations (according to Rinnia) by seeing Rinnia as a girl friend. Apparently Rinnia does not expect a Dutch girl to do that, and this is why she thinks it is 'very nice' of Yvonne that she can get along so well with her.

The Dutch students have no active (dis)like of the migrant students, but mostly seek friendship among themselves. An interethnic friendship is exceptional, as is confirmed by what Rinnia says about her friendship with Yvonne.

*The Dutch on the subject of 'foreigners' in general*

Group discussions were conducted in both classes about 'foreigners' and 'discrimination'. The observations provided an extremely varied picture. Students who at one moment demonstrated a tolerant attitude toward ethnic minority groups were capable of expressing opinions reflecting exactly the opposite a few minutes later. Here are several fragments to illustrate these points. They are taken from a class discussion in which only Dutch students participated. Two of the three ethnic students of this class (class 2) were (by chance?) absent, the third remained silent. The discussion opened with the question:

"Which groups are subjected to discrimination in society?"

The students answered with the following groups: foreigners, the elderly, the handicapped, drug addicts, skin-heads, football supporters, hard rockers, gays and lesbians.

Teacher: "Which are worst?"

Denise and Danny: "Foreigners are worst."

Frida (the girl with the Yugoslav step-father) calls out: "Foreigners are discriminated against because the Dutch are losing their jobs."

Nicole: "They have to leave the country."

Angela: "They have to keep to the rules."

Denise: "Not living with fourteen children in one flat."

Danny: "Nor slaughtering sheep on the doorstep."

Later on, a student reads a text aloud in which someone says, 'those foreigners just breed like rabbits'.

Teacher: "Can you imagine something like that?"

Marjo: "My aunt is always saying, 'those Turks are just breeding to get tax benefits'. She's often saying stupid things like that."

Nicole agrees with what the text says, adding:

"If we did it, they'd kick us out of the country."

Danny, Angela and Denise react to her remark:

"What are you worrying about? You aren't paying for it, are you?"

A bit later on in the discussion – the subject is still discrimination – Denise says that foreigners receive preferential treatment:

"Those types come to Holland and get a house straightaway, and you, you still live with your mother at home because you can't get a house."

Nicole, not surprisingly, agrees, as well as Angela this time. Only Danny reacts angrily with:

"What a load of bullshit."

The only one who is consistent in her opinions is Nicole. She dislikes 'foreigners'. In a subsequent lesson, however, this same Nicole, when asked whether she had ever experienced a situation in which someone was discriminated against , told how, at her previous school, a Surinamese teacher was sworn at and baited by the whole staff. Full of indignation, she relates that other teachers made such remarks as, 'go and do the washing, foreigners like doing that'. When this teacher literally became ill, Nicole and her class took her side. Not exactly the behavior one would expect of someone who dislikes 'foreigners'.

The opinions expressed in the class about 'discrimination against foreigners' could be summarized as follows: 'migrants should not be discriminated against in the sense of being treated unfairly, but neither should they be given preference and they should, just like us, keep to the rules'. The discussions were particularly interesting where opinions sharply differed. Here's how one student reasoned on the subject of 'clothing':
"We all have the right to dress as we like, so they do too." "It's up to them what they wear."
Another student reasoned to the contrary:
"You should adapt to the norms and customs of a country." "If I go to Greece or Turkey on vacation, then I'm not going to go sunbathing topless." 'So "them with their head scarves" should adapt to us here in the Netherlands.'
The discussion in the end turned on the issue of 'how different may they be?' Opinions on what should be kept the same and what may be allowed to differ varied widely, depending on the content given to the concept of 'equality'. The idea, however, that they perhaps had no right to decide on such matters for 'others', in this case migrants, never occurred to the students; they simply took this right for granted.

*Discussion*

The interaction between the Dutch and ethnic students remained generally limited to superficial contact. The Dutch appeared to be less open to inter-ethnic contact than the students of other ethnic origin. They have no outspoken dislike of these fellow students, but neither do they seek their friendship. As Amir stated:
"Making individuals interact across ethnic lines seems to be a major difficulty, because evidence suggests that when given the choice people prefer to interact within rather than between ethnic groups" (Amir, 1976, p. 287).

We see that the negative attitude of Dutch youth to migrants in general is not necessarily reflected in a hostile attitude toward individual migrants in their direct everyday world as secondary school students. Our observations along this line correspond with those of Katz (1986) and Patchen (1982). We further observe a discrepancy between attitudes and behavior: prejudiced adolescents do not necessarily behave in a hostile fashion towards migrants, results which correspond with those discussed in Milner (1983, p. 124 ff) and Stephan (1985).

The attitudes expressed by the Dutch students also provide us with an explanation for the relative caution displayed by the ethnic students in discussing their ethnic background. The specific meaning given by the Dutch students to the notion of 'equality' enables them to wield all sorts of behavioral standards as mechanisms of assimilation. As the standards applied differ from one person to the next, the ethnic students never know to what extent their fellow students will accept differences related to ethnicity. This may result in their attempts to avoid any discussion of such differences. Statements such as "I've been here for 15 years" and "I was born here" can thus be seen in this line of reasoning as strategically made to evade such discussions.

GENERAL CONCLUSIONS

Although practically no one referred to their ethnic origin when confronted with the question, 'Who am I?', ethnic identity appears to be a decisive factor for migrant adolescents when describing the texture of their everyday lives. We see at the secondary schools studied, where ethnics are less than a quarter of the total number of students, that they gravitate together in their social contacts, irrespective of ethnic and gender differences. More specifically, ethnic identification also plays an important role in their choice of friends and (future) partner. The ethnic youth were aware that they are ranked with a group which is socially discriminated against on the basis of appearance or ethnic origin. They have experienced discrimination and know that their opportunities in society are less than those of their fellow students. Also in this sense, their ethnic identity colors their lives and life expectations. At the same time, these youths make assertions to the effect that ethnicity is for them of little importance, and that this irrelevance is morally correct. I have interpreted this as a behavioral mechanism in which a non-discrimination norm is deployed against the discrimination they perceive; a mechanism which, in my view, makes matters worse for them, as it is subsequently difficult to ascribe positive values to (ethnic) differences and differentiations once they have have subscribed to an ideal of equality. This severely limits the room ethnic youths have in which to articulate their ethnic identity. For example, they do select their friends on the basis of ethnic origin, but cannot account for it within the context of their equality ideal. Those who wish to lay claim to equal treatment must, in their eyes, be prepared to assimilate.

The (white) Dutch students expressed a fair amount of indifference with respect to the presence of ethnic students in the class. However, this is not to say that they were 'color-blind' in regarding their fellow students. If that were the case, inter-ethnic friendships would not be so rare and raise so many eye-brows. The Dutch did not express an active dislike of their fellow students, but inter-ethnic interaction remained limited to superficial contact. Some Dutch students did harbor negative opinions toward 'foreigners' in general. Nevertheless, (in class discussions) they rejected any notion of unequal treatment of ethnics. Due to the specific meaning they attributed to the concept of 'equality', they did not consider their own negative opinions about ethnics to be discriminatory: 'ethnics should not be discriminated against, but preferential treatment of them should not be condoned'. 'Preferential treatment' was basically regarded as the opportunity to deviate from social norms, norms which they themselves set as members of the dominant 'white' portion of society. This line of reasoning boils down in the end to a demand for cultural assimilation; in many ways they rejected the ethnicity of ethnic youth. In other words, they may accept their ethnic fellow students as individuals, but not as representatives of an(other) ethnic group.

The ideal of equality harbored by the Dutch and ethnic students alike provides the Dutch with an opening for imposing standards of behavior which, in effect, demand cultural assimilation from their fellow students. Ethnic students, by adhering to the ideal, are deprived of their right to differ, while being subjected to the vagaries of 'Dutch' behavioral standards. As a consequence, any discussion of their ethnic origin may become an uncertain and possibly painful endeavor. In my view, the caution displayed by ethnic students when the subject of ethnic origins is at hand, should be understood in this context.


NOTES

[1] In this paper I shall utilize the term 'ethnic' when referring to those minority groups in the Netherlands whose (ethnic) origins are outside Northern Europe and America; it is obviously not my intention to infer that only these groups have 'ethnicity'.

[2] This study is being conducted in collaboration with Yvonne Leeman of the Centre for Race and Ethnic Studies at the University of Amsterdam.

## 7
# STRUCTURES AND STRATEGIES
# OF DISCOURSE
# AND PREJUDICE

Teun A. van Dijk
University of Amsterdam
The Netherlands

This chapter discusses some theoretical and methodological problems in the study of the relations between ethnic prejudice and its manifestation in discourse. The background to this discussion is a research program which deals with the reproduction of racism in discourse and communication, especially in the context of everyday conversation (Van Dijk, 1984, 1987a), news reports in the press (Van Dijk, 1983, 1988a) and textbooks (Van Dijk, 1987b). The basic assumption underlying this research program is that ethnic prejudices are acquired, shared and legitimated mainly through various kinds of discursive communication among members of the white dominant group. This assumption implies that systematic analyses of discourse about ethnic minorities may provide important insights into two fundamental aspects of racism. First, discourse analysis may tell us something about the content and structure of the cognitive representation of ethnic prejudice, as well as about the properties of their processing during speaking or writing. Secondly, such an analysis allows us to understand exactly how white group members persuasively convey such ethnic prejudices to other ingroup members in communicative interaction and how, thus, ethnic prejudice may spread and become shared within the dominant group.

The analysis of accounts of experiences of racism by black people shows that such characteristics of prejudice in communication hold not only within the ingroup, but also in interaction with minority group members (Essed, 1984, 1988).

Against this background, our study of the expression and communication of ethnic opinions in everyday conversation, based on interview data gathered in Amsterdam and San Diego, has shown that prejudiced white group members generally follow a communicative strategy with two, sometimes conflicting, goals. On the one hand, they positively present themselves as tolerant, non-racist citizens, whereas on the other hand they may (re)present ethnic minority groups in the neighborhood, city or country in negative terms. In this way, internalized social norms of non-discrimination appear to clash with negative personal experiences, or with more general negative attitudes about minorities. In order to resolve what may be both a moral conflict and a practical interaction problem, speakers have recourse to various tactical moves. Such moves typically pair negative remarks about 'foreigners' with assertions which may deny, explain, or otherwise legitimate such negative remarks or their underlying opinions: "I am not a racist but,.."

THEORETICAL FRAMEWORK

The theoretical framework for this study of the expression and communication of ethnic prejudices is complex and interdisciplinary, and will only be summarized here (for details, see Van Dijk, 1987a). Extending the traditional analysis in terms of attitudes to outgroups (Allport, 1954), we analyze ethnic prejudice as a specific type of social cognition, as a negative social representation of ethnic minority groups shared by members of the dominant white group. Such an analysis does not merely specify the content and schematic organization of these social representations, but also their strategic application in ethnic situations (Hamilton, 1981). Prejudice does not consist of the beliefs of individual people, but of generalized opinions shared by people as group members (Tajfel, 1981). This presupposes that prejudice is acquired, used or changed in social situations, and as a function of structures of social dominance. The concrete manifestations of this generalized group prejudice, for instance in individual acts of discrimination, are, however, controlled by so-called 'models' (Van Dijk & Kintsch, 1983; Van Dijk, 1985b). These models are mental representations of personal experiences, for instance, interactions with ethnic minority group members. Under the biasing influence of more general and abstract group representations, members of the dominant group thus build or update ethnic situation models. This may happen in everyday perception or interaction, but also indirectly, through discourse and communication about ethnic events.

Models are organized in a fixed schema, consisting of categories people use to analyze and understand social situations, e.g., Setting (Time, Location, Circumstances), Participants and Event or Action. The propositions stored under these categories characterize not only the personal knowledge people have about a situation, but also subjective, evaluative beliefs, that is, particular opinions. Part of the knowledge and opinions represented in these personal models are instantiations of generalized knowledge scripts, and of (prejudiced) attitudes, respectively. In other words, general group prejudice is tailored to concrete, personal situations through such models. This also explains the familiar finding that everyday talk or action regarding minority groups does not always show ethnic prejudice: Other knowledge or opinions, for instance about the context of interaction or communication, as well as group norms and values, such as tolerance and respect for other people, may effectively block the expression or enactment of such general group prejudice. Thus, whereas shared group representations explain consensus, coherence, and continuity in prejudiced actions of dominant groups, models allow us to explain personal differences and situation specific variation.

One of the typical properties of ethnic information processing is that models of concrete situations are often constructed largely from specific applications of the prejudiced social representations, and not by the information derived from an analysis of a situation with ingroup members as participants. Otherwise neutral events or actions may thus be represented in a biased way, as in the familiar example of the black man sitting on a bench in the park, who may be seen as being lazy instead of enjoying a well-deserved rest from hard work. Conversely, one or a few experiences involving a 'foreigner', once represented negatively in a model, may easily be generalized into a more permanent negative opinion.

Apart from this generalization of models, prejudiced attitudes about outgroups may also be constructed by copying directly prejudiced opinions from existing attitudes about other ethnic groups. This was the case, for instance, for the new immigrant group of Tamil refugees in the Netherlands, in early 1985, of whom the population at large had virtually no experience, and hence no models. Soon, however, the Tamils were attributed properties that were already dominant in prejudices about other minority groups in the country, e.g. "They all want to live off welfare". It is also suggested that talk about minorities is controlled by such ethnic models. This means that biases in the model may also show up in conversation. This is typically the case in stories which white people tell about what they interpret as negative experiences with minority group members. Sometimes, however, for instance in argumentation, such speakers may also express the prejudiced attitude in a more direct way, for instance as generalizations (e.g., "Foreigners are criminals", "Foreigners are favored in housing").

Both model-based and attitude-based statements may, in turn, be controlled by general norms and values, which are also shared group representations. They tell people what they may or should (not) say in specific situations. Again, these general norms and values need to be translated into concrete guidelines for actual (verbal) interaction, and therefore must be specified, in so-called 'context models'. Unlike the models we have discussed above, these models do not represent the situation or events people talk or hear *about*, but the communicative situation in which they are participating. Context models contain information about, e.g. speaker, listener, speech acts and goals. It is this context model that monitors the well-known strategies of impression management or face-keeping. Thus, whereas the situation model of an ethnic event may give rise to negative statements about ethnic group members, the normatively controlled context model of a particular conversation may sometimes block such negative talk, mitigate it, or otherwise transform it into a socially acceptable form. It is thus that ambivalent, but strategically effective expressions, such as "I have nothing against foreigners, but...", arise.

## INTERVIEWS AND DISCOURSE ANALYSIS VS. OTHER METHODS

There are a number of ways in which the non-directive interview and its systematic (discourse) analysis differ from most other forms of experimental testing or field methods of opinion elicitation (see e.g. Hyman, 1975; Plummer, 1983; Spradley, 1979). Conversations in general, but also non-directive interviews, are a more 'natural' way for speakers to express their opinions than responses to pre-formulated questions or the accomplishment of most experimental tasks. Speakers are allowed to specify, explain, correct, or otherwise detail their answers to questions, and may even challenge the presuppositions of leading questions by the interviewer. They may engage in spontaneous expressions of opinion and tend to volunteer arguments or 'evidence', for instance stories about personal experiences, that will make their opinions appear more defensible. At the same time, such informal conversations enable interviewers to disguise their goals, bring up specific topics in a more casual way, or follow special strategies in the elicitation of personal opinions. The same is true in conversations or non-directed interviews on ethnic topics. Although social norms may influence what is being said about minority groups, ethnic opinion will usually manifest itself anyway, if only in an indirect or implicit way. Dialogues have many levels at which such opinions may be expressed, and therefore also be assessed in analysis, for instance in the subtleties of turn taking, semantic moves and presuppostions, lexical choice, syntactic word order, intonation or rhetorical operations, at the local level, or in topic selection and change, and the schematic structures of storytelling or argumentation, at the global level. Some of these characteristics of conversation are not normally under a speaker's control, and may therefore allow more direct inferences about underlying cognitions to be made.

Although some experiments allow unobtrusive assessment of ethnic prejudice, such measurements are only rough approximations of the actual content and structure of prejudice. Systematic analysis of interviews or protocols allows a much more detailed study of the propositional content and organization of underlying cognitions. Ethnic opinions which may be explicitly denied at one point in conversation, may be presupposed or otherwise implicitly expressed or signalled at other points. Repairs, hesitations or pauses may signal doubt or interference with norms, and their analysis may suggest when speakers have recourse to face-keeping strategies. Data from non-directive interviews may sometimes appear to be contradictory, vague or incomplete when compared to forced responses in experiments or questionnaire interviewing and may seem to prohibit precise assessments of underlying cognitions. However, such characteristics of conversation may indeed reflect similar contradictions, vagueness or incompleteness in cognitive representation and processing, including internalized social constraints on the formulation or expression of specific opinions. At the same time, apparent contradictions, both in conversation and in the cognitive representations they manifest, may be made coherent or be resolved at higher levels, involving, for instance, the formulation of different perspectives or points of view of the same event. Adequate discourse analysis can, in principle, handle such complex discursive manifestations of underlying opinions, which usually do not appear in directive interviews or questionnaire responses, and which are seldom analyzed in controlled laboratory experiments.

## PROPERTIES OF DISCOURSE PRODUCTION

Against the background of the general observations made above, we may now discuss in somewhat more detail how discourse structure may be related to the structure of the social cognitions that define ethnic prejudice. Some of these relations may be defined in terms of a theory of discourse production. Although a full-fledged theory of discourse production does not yet exist, its major features are similar to those found in the extensive research on text comprehension (Van Dijk & Kintsch, 1983). An important property of information processing in general, and of discourse comprehension and production in particular, is its strategic nature. For discourse production this means that meanings, words, sentences and various text structures are not generated systematically and precisely according to grammatical or textual rules. Rather, cognitive processing takes place at various levels at the same time, following effective heuristic methods, and using sometimes incomplete information from the current communicative context and from active models, scripts or attitudes represented in memory. Unlike the formal generation of grammatical sentences or well-formed textual structures, this strategic process, especially in spontaneous conversation, receives continuous feedback from the context, other ongoing processes, or representations, and features corrections, trial and error, repairs, reformulations and hesitations.

The overall goal of this mainly unconscious process is to effectively accomplish several things at the same time: the online production of meaningful grammatical utterances, the production of relevant conversational turns, and the accomplishment of speech acts or other social acts, including the communication of information and, ultimately, the (trans)formation of models in the listener. When the speaker has sufficient knowledge of the communicative context, including the actual beliefs and goals of the listener, a provisional production plan may be set up, featuring tentative speech act intentions and relevant 'content' to be conveyed.

The information that is utilized in these strategic processes of production is drawn from various sources. The major source for the 'contents' of the utterance is the current model, representing the event or situation the speaker is talking about. Selected propositions from this model are transformed into locally and globally coherent semantic representations of sentences and texts. The second source is the current context model, which represents the actual goals of the speech participants and other features of verbal interaction. This information will be used for the production of relevant speech acts, for instance an assertion, request or promise. At the same time, thirdly, more general information from relevant knowledge scripts or frames is activated and selectively applied to feed both the situation and context models with necessary inferences. Part of this general knowledge pertains to the structures of sentences and discourse, and provides the rules and special constraints for linguistic production and social interaction. Finally, general attitudes are similarly applied to provide specific evaluative beliefs (opinions) for the situation and context models. These opinions will, for instance, control style, e.g. the selection of specific evaluative words, or the production of specific forms of intonation.

Actual production involves a formulation process which consists of the online production of sequences realizing word, sentence or text forms. These forms embody the strategic expression of underlying semantic representations, signal intended speech acts, and manifest underlying opinions or emotions of the speaker.

It is assumed that all these processes can be effectively coordinated only if they are monitored by a central 'Control System'. This system has many functions in the effective flow of information between short and long term memory, it keeps track of the models or scripts that are activated, allocates processing time to specific sub-processes, represents the current topic or macroproposition of the text under production, as well as the main features of the current context of communication, and generally coordinates the strategies operating at different levels of production.

## COGNITIVE ORGANIZATION AND ITS MANIFESTATION IN DISCOURSE

These highly simplified fundamental assumptions about discourse production are also relevant for the expression of social cognitions such as prejudices. We assume that such prejudices appear in two ways in memory; as general group attitude schemata, stored in semantic or 'social' memory, on the one hand, and as specific situation models, stored in episodic memory, on the other.

Conversations about ethnic minorities may display this dual source. On the one hand, we find formulations of model-based personal experiences, viz., as personal stories. These models not only feature a necessarily subjective, if not biased, representation of earlier events, but also particular opinions, in narrative statements such as "The Turkish family next door makes a lot of noise". On the other hand, we find 'direct' formulations of general knowledge and opinions derived from scripts and attitude schemata, for instance in generic statements like "Foreigners take our houses", or "Turks do not want to learn our language". We find such general opinions in many places of interviews about minorities, for instance in argumentations or in conclusions of stories. These different types of expression in discourse, narrative statements in the past tense versus generic statements in the present tense, suggest a first structural link between discourse structures and the organization of cognitive representations.

However, this simplified picture of the reflection of the cognitive organization of prejudice in discourse obviously needs to be detailed. First, it is highly unlikely that all general propositions expressed by speakers appear 'ready-made' in the scripts or attitudes in social memory. That is, speakers may also express 'new' general propositions. This means that, during production, the expression of general knowledge or beliefs is submitted to an independent process of transformation. General information, like any other information, will be adapted to the requirements of the current communicative context, and may therefore be changed in many ways. Thus, what is generally known or believed about 'foreigners' may be applied to 'Turks' or 'Tamils', even though such information had not previously been stored as such. Hence, we may assume that substitution transformations occur.

Secondly, general statements may also be derived from existing knowledge or attitude propositions by plausible (not logical) inference. If it is believed that "most Surinamese are on welfare", that "many Surinamese drive big cars and dress well", then the general opinion may be derived that "Surinamese get money from illegal sources". This inference may be checked and found consistent with prejudiced opinions about the illegal activities of Surinamese and then admitted to the production process.

Thirdly, general information may appear indirectly by instantiation, that is, the substitution of schema variables by individual constants, e.g., 'Surinamese' by 'our Surinamese neighbors'. Such instantiations are made during model construction or retrieval. This means that, besides representations of actual experiences, speakers may very well express concrete versions of general prejudice opinions, e. g., "Our Surinamese neighbors are on welfare, but have a big car. So they must cheat on welfare and make money illegally". Such propositions may already be resident in the model of the speaker about these neighbors, but may also be locally produced ("come to think of it ...") in specific communicative contexts. These instantiations may lead to particular inferences, which in turn may be generalized as an opinion about the whole group. Thus, the well-known notion of 'generalization' may be explained as general proposition formation in attitude schemata on the basis of a single situation model.

If this analysis is correct, it follows that when general opinions about minorities are expressed in discourse, there does not seem to be an obvious way to establish, for a given speaker, whether these opinion statements are direct formulations of general attitude propositions, whether they have been obtained by various 'on line' transformations, for instance by inference from general prejudices, or whether they are expressions of new generalized model opinions.

There are ways to disentangle these different sources, however. First, general opinions in social memory must have widespread and effective applicability, possibly in a variety of contexts. Cognitive economy will tend to keep general knowledge and attitude schemata as simple as possible: Information that can be inferred from other general information will thus tend to be left out of the schema, unless it is often used in processing. Second, opinion schemata are basically derived from two sources, that is, through abstraction and generalization from models (own experiences, or experiences heard or read about), and directly through communication with other ingroup members or through the mass media. This means that consensual, shared general opinions which are relevant in many communicative contexts and in several kinds of personal experiences (models) tend to be favored in attitude schemata. This may be called the social relevance or functionality principle in the construction of prejudices and of social attitudes in general.

To test these assumptions, comparisons with other interviews and knowledge of public and media discussions about minorities will provide clues about which propositions in an interview are likely to be derived (after possible transformations) from prejudiced opinions in general attitudes, and which are more specific inferences, which may be personal opinions (stored in models) and/or locally produced in the present context. For instance, in one of the stories we analyzed (Van Dijk, 1984), a woman concludes that "Turkish men bring flowers for the doctor, and not for their own wives when these are in hospital". It is highly unlikely that such a specific generalization was previously stored, in contrast to another generalization she makes: "They do not adapt to our rules". Conversational structure will also provide hints about the source and nature of generalized opinions, for instance when a generalization is accompanied by hesitations, false starts or repairs. Finally, ready-made general opinions often have identical formulations, which suggests that they were acquired through communication with other ingroup members rather than through inferences from particular models of personal experiences.

The same reasoning may be applied to the assessment of more specific prejudiced opinions. These are typically stored in, or derived from, models of experience, and will therefore tend to appear in stories about such experiences. Formally, they will be distinguished by names and identifying expressions (instead of variables or generics), by narrative tenses (mostly the past tense), and the semantic categories that also characterize models (time, place, participants, actions, etc.).

We see that there are several properties of discourse which may be considered plausible signals of the mental organization and processing of prejudiced information.

VALIDATION PROBLEMS

Whereas the discussion thus far merely makes plausible guesses about the overall sources and organization of ethnic prejudice and their manifestation in discourse, it still does not show whether specific statements of fact or opinion in interviews may be interpreted as expressions of existing cognitions. In traditional methodological terms, a psychologist may wonder about the validity of the interviews: Do prejudiced statements and their structure express underlying prejudice and its structure, and does the absence of such statements signal that speakers do *not* have a prejudice? In other words, how reliable is such an analysis of prejudiced discourse?

The answer to such questions is complex, and involves many theoretical, methodological and even philosophical assumptions. Before we go into the details of the discourse analysis of ethnic prejudice, let us consider a few general issues. First, assume that a speaker S tells outright lies. For instance, he hates his Turkish neighbor and tells the interviewer that he thinks his Turkish neighbor is "a terrific guy". This, or more subtle, versions of positive self-presentation may occur, but we generally have no way of establishing 'true' opinions in field research, and the same holds in experimental laboratory research. However, especially for extended, non-directive interviews, there are reasons to reject this form of methodological doubt. In interviews such lies would not be isolated. To say the opposite of what one actually knows or believes would require coherent continuation and strategic support, for instance evidence to show why the Turkish neighbor is such a nice guy. Also, such opinions must show consistency with other beliefs about foreigners. In other words, the whole interview in that case should be based on an extensive, locally produced 'fiction'. It is possible that such cases exist, but very few speakers would be able to coherently and consistently sustain such a fiction in a long interview. Therefore, we find it more plausible to assume that, up to a point, speakers in interviews try to tell the truth, even if this will generally be their truth.

In other words, if S hates his Turkish neighbor or Turkish people in general, these opinions will somehow also transpire in talk. It is likely that this will happen more overtly in socially less monitored talk with close friends and family members. However, even with colleagues, relative strangers in public places or, similarly, with interviewers in interviews, such opinions will usually be expressed, either spontaneously or after relevant questioning or priming. However, we suggested that this will take place in an account of *their* version of the truth. That is, negative opinions tend to be mitigated, so as to avoid negative attributions by the interviewer (e.g., "He must be a racist"). This is indeed the case. The analysis of more than 170 interviews, conducted in San Diego and Amsterdam, consistently shows that strong opinions, which may be inferred from other characteristics of conversation, are often formulated in mitigated terms, usually in understatements or similar rhetorical operations, for instance, "Well, I was not particularly crazy about him ... ".

Negative evaluations will often be backed up, either by further arguments in an argumentation structure, or by evidence from models of experience in stories. Such argumentative or narrative support is mainly geared towards the justification of negative opinions, which also contributes to the avoidance of negative attributions by the hearer. In other words, majority group members may have negative opinions about minorities, and will usually show such opinions, but have strategic means to make such opinions appear legitimate or justified.

These strategies of positive self-presentation involve many different moves, such as denials ("I have nothing against them, but ..."), affirmations of exceptions ("You also have good ones among them"), or transfer ("I don't mind, but my neighbors do"). It may be the case that, when taken in isolation, some of these statements may not be (quite) true. That is, speakers saying these things actually *do* "have something against them". This becomes obvious when we analyze such statements in context. First, they are mostly followed by *but*, which shows that there may be exceptions to the general statement.

Secondly, it is this exception which is extensively argued for, or supported by narrative. According to our theoretical model of discourse production, this means that the model features negative opinions. If these are consistent, for the speaker, with general opinions about minorities, then the experience is seen to confirm such general opinions. In that case, it is plausible to assume that the denial of a general negative opinion is indeed a form of positive self-presentation.

If, however, a specific negative opinion, formed on the basis of an unpleasant experience, is indeed inconsistent with (neutral or positive) general opinions about minorities, storytellers will show this in different ways. In the first place, they will avoid telling negative stories about minorities, possibly because they do not spontaneously remember them, because positive attitudes do not facilitate retrieval of negative stories, or else, such negative opinions may not be found relevant for the conversation: Non-prejudiced speakers do not need to make the overall point that they have nothing against foreigners. Rather, such speakers will want to show that their experiences with foreigners have generally been positive. This hypothesis appears to be supported by our interview data. People who may be assumed, on other grounds, to be less prejudiced, not only make positive statements, but spontaneously back these up with stories about pleasant experiences or good relationships with minority neighbors or colleagues on the job. Even when some negative experience does come up, such people will sometimes show at length that this is an exception, or will attribute the experience to other circumstances, and not to properties of minorities as a group. In other words, both positive and negative statements about minorities appear in crucially different discourse structures, depending on whether they are expressed by more or by less prejudiced people.

Thus, positive or negative opinions are seldom expressed alone. There are many rules and structures for discourse and conversation, as well as those for acceptable interaction, which require such statements to be supported, embedded, or explained. It is this complex structure that shows whether or not speakers have prejudiced opinions about minorities. This means that even when something positive or negative is said about an (exceptional) minority person in one part of the interview, other parts of the interview may show that this opinion is indeed dealt with as an exception to the rule.

On the other hand, when prejudiced persons make face-saving positive statements about minorities, or deny being prejudiced against
them, they should not simply be assumed to be lying. This may be
explained as follows. First, such statements may express general
norms or values that the speakers may support in general (Billig,
1988). They thereby show that they respect the basic norms of society, and that therefore they see themselves as normal, that is, non-
racist citizens. For them, negative opinions expressed about minorities are not inconsistent with this self-image, simply because a legitimate complaint is not seen as evidence of a racist attitude. Indeed, to
resent assumed abuse of social welfare is in agreement with other
norms and values of society, so if it is observed or believed that
some foreigners do indulge in such abuse, it is found legitimate to
have negative opinions about this. Even the (unjustified) generalization from such negative models of experience may be thought to
hold if selective evidence may be produced which supports such
claims. The media, or communication with others, may be mentioned as such support, even if sometimes prefaced by the strategic
statement "We may of course not generalize, but...". It is the task of
a cognitively oriented discourse analysis to reconstruct this biased
version of reality, by relating opinions expressed in interviews to
other underlying opinions, norms and values.

These observations suggest that there is considerable theoretical and
methodological evidence to take interview discourse at face value.
Despite many due to communicative and interaction strategy goals,
we assume that accounts of experiences, as well as general statements, do indeed reflect what speakers actually think. However,
statements cannot be interpreted in isolation. Only the complex
structure of the discourse can be related to the structures and strategies of cognitive representations and their processing. Therefore, we
will show in a last section, in somewhat more detail how such discourse structures and strategies may be interpreted as empirical evidence for cognitive structures and strategies.

STRUCTURES AND STRATEGIES OF DISCOURSE AND PREJUDICE

It was assumed above that the theoretical distinction between prejudiced attitude schemata and prejudiced models may also be
observed in discourse, for instance in different discourse structures

or genres, such as stories and argumentations. The next step in establishing this correspondence between talk and thought is a further analysis of more detailed structures of attitude schemata and models and a comparison with possible correlates in discourse, or vice versa.

## Structures of prejudice

In our theoretical framework, attitudes are defined as hierarchical structures in semantic (social) memory, consisting of general, evaluative beliefs, that is, general opinions (see also Abelson, 1976). In that perspective, prejudices were taken to be negative attitudes shared by members of a dominant group about a dominated outgroup (Allport, 1954). This does not mean that each opinion in such a structure is negative, but that the higher-level, controlling macropropositions are negative. For effective storage, access and retrieval, opinions in prejudiced attitudes are organized by labeled nodes, or categories, such as Origin, Appearance, Cultural characteristics or Personal features. These semantic categories organize opinions in similar domains of experience or observation. Together, these categories form a hierarchical schema, in such a way that the general opinions which may be inserted in their 'slots' also show different hierarchical positions. For instance, the proposition "Minorities are criminal" is more specific than the overall macroproposition "Minorities have negative character traits", but more general than the lower level proposition "Minorities are involved in the drug business", which again is more general than the prejudiced opinion "Surinamese men are often drug pushers". The same is true for the organization of opinions about origin, appearance, cultural habits ("They do not speak our language", or socio-economic competition ("They take our jobs", "They take our houses").

The formation, .acquisition, and application of ethnic attitudes is partly determined by such organizing categories, which may be assumed to form a fixed 'prejudice schema'. Hierarchical structures of this schema facilitate fast access to high level opinions. Categorization of opinions allows selective addressing and retrieval of specific opinions, e.g. those of competition, or even those of competition in housing. Such selections may be structural or ad hoc (contextual). Typically, people who are esspecially concerned with competition in employment, e.g. because they are unemployed, or because they have minority competitors or superiors on the job, will tend to focus on this category. They may even form or transform their attitude schema in such a way that such a category may be placed at a higher level, and become more important than negative opinions about, e.g. different appearance. In other words, the prejudice schema formed by these categories also defines what information about minorities is most relevant for different subgroups of the dominant group.

This assumption also allows flexible integration of the theory with a theory of class or dominant white group factions: Different experience and different social position lead to differently organized ethnic prejudice, and partly to different prejudice content. This important dimension of prejudice structures and their (trans)formation will not, however, be further discussed here, and belongs to the (much needed) sociological component of a theory of social cognition (see also Van Dijk, 1988b).

*Prominence in discourse*

The structures of prejudice as postulated above will, at least in part, also show in behavior, and hence also in communication and conversation. In simple terms, a first hypothesis would run as follows: What is prominent in the prejudice structure will also tend to be prominent in speech. That is, all other things being equal, people prefer to speak about what they have on their minds. This principle, which is related to fundamental principles of memory search and retrieval, and which is often studied in terms of 'availability', also has consequences for discourse structure and communication.

Thus, when people are asked, for instance, about their neighborhood, the question will first trigger and retrieve high level knowledge and beliefs about the neighborhood. Very common in our inner city interviews, for instance, is the general opinion that 'the neighborhood is run down', or that 'they do not like it'. A following why-question by the interviewer, or spontaneous follow-up of the interviewee, will then trigger high level reasons, as represented in the model people have about their neighborhood. Often, such reasons will feature concepts such as 'foreigners' or 'immigration' as a major element in attribution and explanation. This is also the case in the 'fundamental attribution error' in the explanation of actions of outgroup members (Pettigrew, 1979). Whereas failure of self or ingroup members tends to be attributed to circumstances beyond our control rather than to personal characteristics, the opposite is the case for negatively valued actions of minority groups. That is, minorities are often personally blamed for structural inequalities in society of which they are victims: Instead of miserable educational facilities, their 'lack of motivation' is seen as the major cause for educational 'underachievement'. Similarly, in our example, the combination of concepts or propositions featuring 'neighborhood', 'run down', 'crime' or 'housing' will tend to trigger high level information in the 'personal characteristics' category of the prejudice schema about foreigners, rather than structural factors of urban decay in other scripts or attitudes. In other words, the ordering of explanatory topics spontaneously brought up in conversation gives us clues about the hierarchical and categorial organization of prejudice schemata and models, as well as about the perspective of the speaker in the establishment of causality (see also Taylor & Fiske, 1978).

Note that this assumption does not beg the question: We do not simply assume that prejudiced opinions are prominent because they are prominent in discourse, and then conclude that prominence of prejudice leads to prominence in discourse. This correspondence is independently established on theoretical grounds, such as the principles of search and retrieval or cognitive organization. Similarly, the correspondence may also be corroborated by other discourse structures: Prominence may also be signalled by the length of, or the amount of detail in stories, as well as by speed of response, or lack of search pauses. Data from other research (experiments, surveys), and especially significant convergence of the discourse structures of different interviewees, may provide further independent evidence for this correspondence between discourse structure and prejudice. When asked a general question, respondents very seldom begin with a specific story. They first produce general statements, which in turn may trigger more specific models of personal experience. Cognitive strategies in this case may be similar to, and even determine, conversational ones relating to topic introduction, level of description, and topic shifts.

## Models and stories

In similar ways, we may try to find discursive evidence for the structure and content of mental models. We have repeatedly suggested that models tend to be expressed as stories: Models represent situated, personal experiences, featuring specific Time, Place, Circumstances, Participants and Actions (for details, see Van Dijk, 1985b). These categories of model schemas may appear in semantic representations of propositions and their syntactic expressions in sentences (Dik, 1978). Secondly, these categories also appear in the conventional structural categories of stories (Labov, 1972).

Our field work has shown that stories about minorities have rather a homogeneous structure and content. Firstly, they are usually embedded in argumentations. They are not primarily told to amuse the hearer, but to make a point, to provide evidence for a general opinion. This opinion statement usually precedes or follows the story and is explicitly linked to it. This characteristic of storytelling in the interviews may be interpreted in cognitive terms by saying that models are primarily triggered, retrieved and actualized as a function of general opinions, that is as an 'illustration' or 'proof' of credibility or legitimacy.

Secondly, the Orientation category of these stories tells us something about the nature of the models in which ethnic minorities tend to appear as participants: They are descriptions of mundane everyday activities, in which the storyteller or another protagonist is 'simply going about his/her daily business'; shopping, taking a walk, or working. Cognitively, this suggests a model structure in which an innocent I is represented, and contrasted with the guilty villain (the foreigner). Expressed innocence and mundanity also have a narrative function: It makes the following Complication category of the story more prominent, exceptional and unexpected. Similarly, the contrast between innocent speakers, represented as victims, and foreign villains is, of course, also an important narrative strategy to persuade in order to enhance credibility and get sympathy from the hearer. This cognitive structure is consistent with more general prejudiced opinions, in which the ingroup is represented as the victim of immigration, and the outgroup as people who violate 'our' norms, habits, safety, privileges, and even the routines of our everyday life. Thus, general attitude structure may guide the organization of specific model structures, which in turn will appear in specific story structures.

Next, most Complications in stories are about negative actions of minority group members (assaults, stealing, everyday nuisance and harassment, dirtiness, cheating, etc.). Note that, in principle, an infinite number of specific stories may be told about minorities. This is not the case, however. Such stories have a very limited topical repertoire. Most actions described in these stories are instantiations of general, prejudiced opinions. That is, the stories of personal experiences may be about personal experiences, but the storytellers select surprisingly similar experiences to relate, and these happen to be very close to the stories read or heard about in the media, or in everyday conversation. In other words, there are not only cognitive stereotypes (attitudes and models), but also narrative ones, which through further conversation may of course contribute to similar models in other group members.

Unlike other stories, these stories often do not have a Resolution category. Again, this is in line with what may be predicted from the underlying model structures: Resolutions usually feature heroic, courageous or lucky acts of protagonists, defeat of opponents, or solution of a problem. For prejudiced speakers, none of these can be the case. Because they see and represent themselves as victims, and not as heroes, the villains continue to be a problem for them. In other words, there is no solution to what they see as the foreigner problem, and this is probably also how ethnic situations are represented in their models. Despite the formal constraints of narrative structure, storytellers therefore will tend to omit the Resolution category. They tell what may be called a 'complaint story', which focuses on the Complication rather than on the Resolution.

Finally, the Evaluation category of these stories may be interpreted as the expression of the opinions represented in ethnic situation models, whereas the Coda or Conclusion category, usually expressed in sentences with verbs in the present tense, exhibit either current plans of the speakers ("I won't go out at night anymore") or more general ethnic opinions that the story aimed to illustrate or prove (e.g., "You can't trust them").

*Other discourse structures*

Other structures of discourse may also be related to cognitive representations and strategies of prejudice. Our analyses of a large number of interviews have suggested the following links, which will be summarized in a few points.

1 Semantic structures realize communicatively relevant fragments of both models and general prejudices. These represent in discourse what people think about minorities and ethnic relations. Semantic analysis, however, should be made relative to the overall meaning structures of the interview, not of propositions in isolation. Conceptually, these are organized in a few basic categories: Difference (of appearance, culture and behavior), Deviance (of norms and values, e.g. in crime), and Competition (for space, housing, jobs, education and welfare). Along another dimension, these semantic structures may also be summarized in terms of the notion of (perceived) Threat.

2 Propositional structures of sentences, as well as their syntactic formulation, for instance in the description of action, reflect the underlying perspective or point of view, as represented in model structures. In this way, agents responsible for negative actions, viz., minority group members, tend to be given a prominent, initial position, that is, as syntactic subjects ("They deal in drugs"). In this way their agency and responsibility is made more prominent than if they had been mentioned in the downgraded prepositional phrases of passive sentences ("Drugs are sold by them"), or if they had been left out completely ("There is a lot of drug dealing going on"). This is particularly the case in written language, for instance that of the media (Fowler, *et al.*, 1979; Van Dijk, 1988a).

3 We have seen that prejudiced discourse often features disclaimers of various types, e.g. mitigations, denials and apparent concessions. The analysis of these semantic or rhetorical moves is also relevant in relating discourse structures and cognitive structures. The general pattern of these disclaimers, viz. 'A, BUT B', where A is a positive self-description and B a negative other-description, suggests possible conflicts between the content of situation models on the one hand, and context models on the other. This conflict is both interactionally and cognitively resolved by having recourse to such face-saving moves, thus realizing essential steps in the overall strategy of impression management. More specifically, this suggests how norms and values, embodied in the representation of self in the context model, control which information

from the situation model must be selected or modified in talk. Hence, disclaimers or other semantic moves may be analyzed as signals of underlying socio-cognitive conflicts, and their strategic resolution, when prejudiced people discuss a delicate topic such as foreigners (see also Billig, 1988; Potter & Wetherell, 1987; 1988).

4 The stylistic level of lexical choice and syntactic variation does not directly express specific cognitive content, but signals perspective and opinion, as well as social constraints of the communicative situation, such as the degree of formality or familiarity of language use, and the status or group membership of the speech partners. In prejudiced descriptions of minorities, for instance, we witness a general reluctance to use names as identifying expressions, a preference for excessive pronominalization and the use of demonstratives ("they", "those people"). Such expressions may be interpreted as signalling social distance, and therefore also exhibit some of the characteristics of the models people have about themselves and their relations with minority groups and their members.

5 Rhetorical structures are typically oriented to the communicative context. They serve persuasive functions, i.e. to emphasize specific content, points of view or opinions. Such communicative functions not only express the underlying structures of communicative situation models, e.g., what the speaker thinks about the hearer, but also what is most relevant, important or otherwise remarkable for the speaker. Thus, the opposition between 'us' and 'them', prevalent in both models and attitudes, may rhetorically be enhanced by contrast, as in *"we always work hard, and they can have nice parties every week"*.

6 Finally, specific conversational properties of spontaneous speech, such as turn-taking moves, repairs, false starts, pauses, or variations in intonation, stress or volume, not only have interactional functions, (for instance of face keeping, credibility enhancement or persuasion), but they may also signal subtle properties of cognitive operations during discourse production. Thus, search and retrieval of relevant models and attitudes and their contents, the construction of semantic representations, the selection of appropriate lexical items and the final formulation in grammatical structures all require complex strategic operations. Positive self-

presentation imposes a high level of self-monitoring, which will lead to heightened control of exactly what is said, and how it is said, especially with delicate topics such as race relations. This will probably cause more pauses, false starts, repairs and similar manifestations of the cognitive strategies of optimal expression of meaning (see also Levelt, 1983). We have found, for instance, that not only do prejudiced people tend to make prominent use of referential expressions which denote ethnic minorities, but they also often pause, hesitate or make false starts as soon as they must name or describe minority groups or their members. It seems likely that these surface phenomena signal the speakers' awareness that the choice of the 'correct' term is important in naming ethnic minority groups.


CONCLUSIONS

The fact that non-directive interviewing is a powerful methodological instrument in the social sciences, has been demonstrated in many earlier studies which emphasize the importance of personal accounts of people's experiences and opinions (Harré & Secord, 1972). When such accounts are given by members of a dominant group, and deal with their relationships with a dominated group, they may often be expected to be self-serving, biased or even prejudiced. Our work on the reproduction of racism in discourse, and in particular our analyses of interviews which resemble informal everyday conversation, suggests that such accounts provide crucial data, which cannot be gathered by other methods.

From a sociological point of view, the question of the validity or reliability of such subjective accounts, when compared to what people really think or what really happened in the episodes talked about, may be less relevant than the question of how people actually and observably go about the mundane but delicate task of talking about minorities. Also, whatever such speakers, as group members, actually think, it is what they tell others which is relevant for the expression of ingroup membership, for intra-group communication, for the construction of an ethnic consensus and for the confirmation of group position in a multi-ethnic society.

From the point of view of social psychology, however, the analysis provides important insights into the representations and strategies involved in ethnic prejudice. Besides the usual experimental methods and the field surveys which attempt to assess the contents and the relevance of ethnic prejudice, such accounts yield very rich data for a study of the social cognitions of ingroup members in the study of intergroup relations.

Our research suggests that there are multiple links between the content, structures and strategies of such cognitions, on the one hand, and those of the discourses that express them, on the other. Despite many types of transformation, to be accounted for in theoretical terms of discourse production and within the perspective of rules and other constraints of social interaction, discourse in many ways shows what goes on in the minds of people. In contrast to brief responses in structured interviews or questionnaires or to the often unnatural tasks in laboratory experiments, non-directive interviews provide an optimal way of eliciting experiences and opinions of a delicate nature. We have suggested that, in such a complex context of communication, people will not, or even cannot, consistently dissimulate about their experiences or opinions. Given the occasion to speak their minds about what they see as their problem, they will also give their version of the truth.

Against this methodological and theoretical background, we are able to engage in a systematic study of both ends of the relationship: We may search for the expression in discourse of postulated structures or strategies of the social cognitions of prejudice and, conversely, we may try to explain typical and recurrent features of talk about minorities in terms of the structure of ethnic opinions. It is important, however, that such an analysis goes beyond the traditional methods of superficial content analysis, or even of those of contemporary protocol analysis in psychology. A highly sophisticated and subtle discourse analysis is necessary to trace and describe the many ways in which dominant group members show their underlying opinions (see the contributions in Van Dijk, 1985a, for further theory and analytic methods).

A brief summary of some research results has shown that such discourse analyses are viable and yield multiple new insights into both prejudiced talk and prejudiced opinions. Thus, argumentation structures can be linked with general attitude schemata, stories with episodic model structures of personal experiences, thematic structures with high-level hierarchical propositions of both models and attitudes, and local semantic moves with both the conflicting goals of positive self-presentation and negative other-presentation, goals which are represented in the underlying model of the communicative situation. Similarly, semantic structures of propositions, as well as their syntactic expressions, may signal a point of view on ethnic events, whereas lexical style will invariably manifest both communicative constraints and opinions. Rhetorical and conversational structures function within the interactional context, for instance in order to heighten credibility and enhance persuasive impact, but also show subtle underlying structures (e.g., the opposition between 'us' and 'them'). Similarly, they signal some details of production processes which are difficult to assess with other methods, as is the case for pauses, hesitations, false starts and repairs, which may be related to search, retrieval and cognitive decision procedures during the formulation of descriptions and opinions on delicate subjects in the most appropriate way.

We see that discourse analysis, as an interdisciplinary field of inquiry and with a set of highly sophisticated methods, may in principle contribute new insights and open up new directions of research in the social psychological study of ethnic prejudice and the reproduction of racism in society. In order to gain an insight into the social cognitions of dominant group members, and provided that we use a serious theoretical and analytical approach to such personal and social accounts, we may indeed follow the elementary principle "Why not ask them?"

# 8

# THE DEVELOPMENT OF ETHNIC TOLERANCE IN AN INNER CITY AREA WITH LARGE NUMBERS OF IMMIGRANTS

Wiebe de Jong
Erasmus UniversityRotterdam
The Netherlands

For more than twenty years I have lived in an old working class area in the centre of Rotterdam, which in 1969 was still designated by the City Council for demolition for the purpose of urban expansion. Since 1976, however, urban renewal has been in full swing, with the result that, in 1985, 45% of the residents were living in new dwellings. In 1985, almost half of the 9500 residents were immigrants. These people came from various ethnic backgrounds: from Surinam, Turkey, Morocco, Cape Verde, Spain, Portugal, Yugoslavia and China. Almost all residents, including the Dutch, belong to the lowest socio-economic strata. In 1983, 39% of the work force was unemployed (all ethnic groups suffering about equally).

I soon became actively involved in a wide range of activities in the area: education, help for immigrants, opposition to the imminent demolition of the area. I thus met many people of different ethnic origin. This enabled me to follow closely the development of inter-ethnic relations between 1970 and 1985. When I decided to systematically analyze these experiences, I became aware that being actively involved in the development of the area made me an 'interested party', which might bias my perceptions. For this reason, I have, in addition to my own experiences, made use of external sources whenever possible. Fortunately, at a number of locations in the area – some of them crucial to my research project – records (minutes, letters, memoranda, work group reports, etc.) had been conscientiously kept. This was my first important external source. I have also interviewed a large number of people who were key figures in certain situations, and have studied the official documents relating to the area. In addition, I have consulted a large number of articles about the area which have been published over the years in daily and weekly newspapers and professional journals. To this I have added a statistical analysis of the number of members of various ethnic groups who have come to live in the new estates of the area.

The material I have collected relates to processes of decision making, actions, definitions of the situation and dealing with problems as they have occurred, at group and intergroup level, and to their impact on the behavior of members of the various ethnic groups. My analysis deals with the group and intergroup aspects of the development of ethnic relations. I did not go into the question of people's individual experiences in this field.

POINTS OF DEPARTURE FOR THE RESEARCH PROJECT

My intention was to describe interethnic relations in this area as completely as possible. Most suitable to this end are descriptions at various levels. It seemed best not to choose one single theory of interethnic rations, nor to limit myself to only a few variables, but to take into account as many variables as possible. This, after all, provides the opportunity to explore, at any given moment in the development of interethnic relations in the area, which variables were of importance.

As dependent variable in this research project, I have taken a behavioral concept: *ethnic tolerance*. Ethnic tolerance is tolerance towards people of different ethnic origin in one's own surroundings (e.g. residential and work situations) such that ethnic origin does not constitute a reason for withholding rights, positions and opportunities. Tolerance does not necessarily mean liking each other or relating to each other as friends. One may continue to prefer having intimate relations within 'the own circle'; one may continue to see different cultural customs, role patterns, standards and convictions as foreign, and still tolerate one another in the residential situation and, for example, in the joint representation of interests resulting from that shared residential situation.

Tolerance alone is not enough. In South Africa, the blacks are 'tolerated' as a pool of second-class workers or domestic helps, and in certain areas in The Netherlands, ethnic minorities are 'tolerated' in slum dwellings, while the indigenous population lives in new dwellings. In both examples it is the ethnic origin which determines the second-class position. In choosing a behavioral concept as a dependent variable, I did not assume a behavioral explanatory model. The same behavior can be caused by very different factors. In this research project, I have found that tolerant behavior towards ethnic minorities may result from totally different motives:

- One may have *positive attitudes* towards people of different ethnic origin.
- People may have a *normative conviction* that ethnic origin ought not to be a reason for withholding rights, positions and opportunities. This constitutes an active adherence to values and may be quite separate from whether or not one likes members of the group in question.
- People may be guided by *reference groups* (and/or reference persons) who express the view that ethnic origin ought not to be a reason for withholding rights, positions and opportunities. If people are ethnically tolerant because they are taking their lead from these reference groups of persons, it entails an active personal commitment to that behavior.
- Ethnic tolerance may also arise as a result of *mere behavioral conformity* with the groups to which one belongs, for example, the norm prevailing in the group that mutual tolerance is necessary, This norm is then adhered to in order to avoid being excluded from the group, although privately the norm is not subscribed to. It is thus a case of passively conforming to the prevailing norm within the group.
- People may also accept the norm of ethnic tolerance passively because they are afraid that otherwise *their interests* will suffer.
- People may also be *powerless* to put up resistance against the granting of equal rights, positions and opportunities to members of ethnic minorities.

There will be people who have a positive attitude towards all other ethnic groups. Others will at all times attach overriding importance to their conviction that people should enjoy the same rights and opportunities, regardless of ethnic origin, and there are people who will never reconcile themselves to such a view (prejudiced individuals). Often, however, people will not be influenced by only one primary motive, but by a combination of motives, e.g. a combination of cultural values and convictions adhered to by reference groups which are important to them, passive acceptance of the norms of groups or institutions, and their own interpretation based on their social situation and on relationships within the specific situation.

It is important, both in the theoretical and practical spheres, to understand that various motives may underlie ethnic tolerance. If people's motives are affected not only by personal factors (such as having positive attitudes and/or values) but also by intragroup and intergroup factors, and structural and cultural factors, it means that, in order for us to be able to analyze ethnic tolerance, it is not enough to find out what people think about it. We must also ascertain how the situation is defined within the groups to which individuals belong, what the state of intergroup relations is, and how structural and cultural factors in the district concerned, and in society at large, affect them. At the same time, interactions between all these factors must also be considered. In practical terms, these theoretical principles mean that there is no universal road to ethnic tolerance: attempts to initiate a process leading towards it may be undertaken at various levels and in various ways. I should like to illustrate this by means of a concise account of the development of interethnic relations in the district known as 'the Old West' between 1970 and 1985.

DESCRIPTION OF THE SITUATION

At the end of the sixties, Dutch industries took on large numbers of foreign workers, especially from Turkey and Morocco. As a result of the housing policy of the city of Rotterdam and the policy of housing corporations, foreign workers were forced to settle in the old areas like the district 'the Old West', studied in this paper. This was where cheap, but run-down rental accommodation could be found. At this same time, the City Council published a memorandum on renovation, announcing the demolition – in phases – of these very areas. As a result of municipal policy and the housing situation which arose in the Old West in the period 1970-1973, many of the indigenous population felt that they were being forced into competition with the immigrant workers. As cheap rental accommodation was so scarce in Rotterdam, it was impossible for the original residents of the area to move away en masse, which made them continue to draw on the houses available in the district (e.g. for the benefit of their married children), but many landlords preferred their tenants to be immigrant workers, as they made fewer demands. This competitive situation, however, was not the only reason for the strong feeling against the ethnic minorities. Many members of the original population also felt insecure because their basic convictions were being threatened. Uncertainty about the future housing situation was only one factor. At this time, in the interactional sphere, the original population were suddenly confronted by large numbers of Turks and Moroccans--mainly single men--whose mores and views on role patterns were very different form those prevailing in the area, and who, moreover, did not speak Dutch. The housing arrangements for immigrant workers in rental accommodation converted into boardinghouses often prevented contact between the communities. This reinforced the perceived differences between them and gave rise to outgroup dislike.

Due to this combination of factors in the Old West, members of ethnic minority groups did not encounter a friendly reception in those years, in contrast to the findings of other researchers in The Netherlands (e.g. Bovenkerk *et al.*, 1985). In the Old West, the intolerance among the indigenous residents was so strong that even a group which was important to them, i.e. an action group, did not succeed in putting across to the residents its definition of the situation (that not the immigrant workers, but government and industry were to blame for the fact that immigrant workers came to live only in the old districts). This action group was set upon in 1970 to fight against poor living and housing conditions and the impending demolition of the area. It was a voluntary organization which made decisions and planned actions in fortnightly meetings. The leaders of the action group were 'compelled' by the residents to adopt squatting as a course of action: taking unauthorized possession of vacant houses for the benefit of Dutch families. Thereafter, because the municipal authorities did nothing to spread the immigrant workers over the entire city, this policy, which had initially served a warning function, was gradually adopted as an action against the immigrant workers who wished to settle in the district.

The reason the major race riots in Rotterdam, which hit the national headlines in 1972, broke out, not in the Old West (as large sections of the media had predicted), but in another part of Rotterdam, may have lain with the fact that nothing happened in the Old West to trigger off a riot. It does not take much to provide the spark, and in the Old West as well certain situations could surely have provided a suitable opportunity. A more likely explanation is, perhaps, that the squatting campaign carried out by the action group acted as a safety valve for residents' frustrations. The way in which the action group operated (with fortnightly meetings, regular campaigns to keep the area residential, and so on) gave the Dutch residents the chance to identify with this organization. This may have had the effect that people were inclined to turn to this organization before taking measures of their own. On the other hand, the same organization reinforced the idea among the indigenous population that it was impossible to live with other ethnic groups, a view which was expressed in deeds as well as words. In this period, therefore, there was no ethnic tolerance on the part of the indigenous residents and the organization representing their interests.

In the period 1973-1976 it could also not be said that ethnic tolerance prevailed among the indigenous population or in the action group. However, the configuration of factors which led to this situation was not the same as it had been in previous years. Until 1974, the municipal authorities adhered to their policy of demolition of the area, but within a relatively short period of time after that, the authorities changed their view of what should be done with old areas: urban renewal instead of demolition. However, the results of this were not yet visible in the Old West, as no concrete building activities had yet taken place. The exodus of the Dutch residents continued at an increased rate, because the latter--more than in the preceding period--had an opportunity to obtain better rental accommodation in other areas. This combination of factors gave added impetus to the process of segregation which Rex (1981, pp. 25-42) regards as characteristic of situations in which immigrant groups take over an area from the indigenous residents. However, this process did not result in one group taking the place of another: rather, it looked as if the area might become "a melting pot for population groups who had nowhere else to go". In January 1975, the Old West had a population of 11,182; of these, 2,305 (20.6%) were of foreign nationality. There were 776 Turks, 436 Spaniards, 287 Moroccans, 284 Yugoslavs, 253 Portuguese and 269 people of other origin, including Cape Verdians, Italians, Greek, Chinese, etc. In addition to this 20.6%, a further 8% were from the former colony of Surinam.

The situation was not the same as it had been during the earlier period, as there was no longer direct competition for housing between the indigenous population and the immigrant population: the indigenous population was moving out. At this time, therefore, ethnic intolerance on the part of the indigenous population was caused less by competition than by a sense of threat, because of growing uncertainty. Local people were still uncertain where they would be living in the future. Moreover, various networks which had previously existed were destroyed because neighbours and relations were moving away. Their place in the district was taken by immigrant families, for it was in this period that family reunification among immigrants started. Immigrants, therefore, resorted less and less to boarding-houses and more and more to the usual rental accommodation. Thus, more and more Dutch residents now had immigrants living next door, with whom they were unable to speak and whose habits were very different. In the living situation as it existed in the Old West at the time, the fact that people lived close to one another, and that there was a certain interdependence, did not lead to a greater understanding among the Dutch residents for the immigrant population. On the contrary, it was a process in which the idea one had of immigrants 'being different' was reinforced, because Dutch residents – as Thibaut and Kelly (1959) and Amir (1976, p. 288) state – could see for themselves that various 'alleged' inequalities were genuine inequalities, e.g. the specific role patterns for men and women, education, and other mores, standards and values.

At meetings of the action group, which were attended only by Dutch residents, the feeling of being threatened was frequently expressed. In contrast to the first period, however, the action group no longer took action against immigrants. Members ceased to concentrate on the past and on present complaints and concerned themselves more with the future and ways in which to improve the area and its housing. Discussions about 'the immigrants' were regarded as being of secondary importance to this aim. However, the area's Dutch residents were taken as the reference group for future plans for the area. What they wanted was a 'new' district with a balanced population. Its nucleus was to consist of Dutch families. These ideas were not formulated explicitly in this way, because it was presumed that immigrants would eventually return to their native countries. When reference was made to interests, rights and opportunities, what was meant were the interests, rights and opportunities of the Dutch residents. It was considered acceptable for as many immigrants as there was room for to live in the area.

In the period of *1970-1980* the process of urban renewal continued. Building was in process in all the old areas of Rotterdam. There were great differences in population between the areas. In a number of areas, the new houses were almost all occupied by Dutch residents (Smit & Noort, 1987). This was not the case in the Old west. How does one explain these differences? An important factor was the fact that urban renewal in Rotterdam was based on a policy of decentralization. In project groups in the areas, mandated officials and residents could make decisions on the planning, execution and future occupancy of the new estate. This resulted in different accents in different areas. In the Old West, building activities clearly resulted in fewer Dutch families moving out of the district than previously was the case. The Dutch residents who remained in the area now regarded the immigrants as competitors for the *new* homes which were being built. This fear was expressed at meetings of the action group.

It was true that the action group continued to concentrate fully on a 'new district', but it was no longer possible to evade 'the immigrant question', as it had been in the previous period. Decisions had to be taken about the rules for allocating the new homes. It was only possible for new homes to be built if old ones were first demolished: there were no vacant lots or open spaces in the area, as opposed to other old Rotterdam districts where new dwellings could be built on large vacant lots. That was why re-housing and occupancy of the new estate were closely interwoven in the Old West. Consequently – on technical grounds – the leaders of the action group redefined what many Dutch residents regarded as a situation in which competition prevailed: they decided to call it "a situation of mutual dependence", for, in order to be able to replace all the old housing with new homes without delay, it would be necessary for everyone to cooperate, both the indigenous and the immigrant populations. The action group also made it clear, however, that the indigenous residents would not be the losers in this situation, as they considered it unlikely that immigrants would be very interested in new housing. As a result, the 'objective housing allocation rules' were adopted in 1970 after a year of debate in meetings of action groups. Regardless of how long they had been living in the area, and regardless of origin, those inhabitants whose dwellings were destined for demolition for the purpose of district renewal were first eligible for new housing.

Although the economic crisis in The Netherlands had a very perceptible impact on the Old West (in 1983, 39% of the professional population was unemployed, as we have seen), we nevertheless find, in the period *1980-1985*, an increasing ethnic tolerance with regard to the level of housing, representation of interests, and facilities. The fact that economic crisis may play a negative role in the degree of people's ethnic tolerance seems to be well documented (Van Niekerk, Sunier, & Vermeulen, 1987), but this need not be an all-determining factor. In my opinion, this is dependent not only on the degree of continuing economic and social security, but also, as in the district in which I did my research, on the way in which the housing situation is arranged for indigenous residents and immigrants, and on the way in which intra- and intergroup factors influence the relation between the various groups. I particularly want to explore the intra- and intergroup factors which were important for the emergence of ethnic tolerance.

When the action group formulated the 'objective' housing allocation rules, an important prerequisite for the emergence of ethnic tolerance in the Old West was created, but as long as the action group still formulated ideas in terms of 'indigenous versus immigrant' in mind, and as long as the indigenous population were seen as residents and the immigrants as foreigners, actual attainment of equal rights and position would not be possible. After all, immigrants did not yet have the right to make decisions and they did not yet make use of the opportunity to live in the new houses. As such, the housing allocation rules were certainly a condition necessary for the inception of ethnic tolerance, but not a condition sufficient in itself. Before this tolerance could be attained, a different definition of the

situation was needed, and, in conjunction with it, an altered definition of the problem. The fact that it was possible for these alterations
to take place was due to an influential group of volunteers within
the action group, who succeeded in making their definition of the
situation the basis for the future policy. This group, known as the
Educational Advice Centre had, in choosing which projects to initiate (orientation classes, homework classes, an international children's library, language courses for foreign parents) aimed at the
needs of all parents and children in the area, regardless of ethnic origin. After heated discussion between members of the action group,
it was decided to emphasize that everyone in the area should accept
the fact that immigrants were residents of the area, that they had as
many rights as the Dutch residents and that the different ethnic
communities should cooperate in dealing with common problems.
With this, the action group set a norm of mutual tolerance.

Whether or not one feels bound to the norms which a group sets is
determined by the importance of the group. This was true for the
action group. Thus, the new policy did not cause the Dutch residents, many of whom initially had different ideas about the issue, to
sever their links with the action group. Why did this organization
remain important to the indigenous population? Firstly, it brought
about urban renewal. Secondly, the importance of the action group
as an organization was greatly increased by the leading role it
played in an effective anti-heroin campaign. Since 1975, the Old
West had increasingly become the centre for heroin trafficking in the
Rotterdam area; this was partly because of its central location and
partly because many buildings were temporarily unoccupied during
the renewal process. An additional complication was the fact that
Surinamese creoles had a monopoly on street-trafficking. Actions
against the heroin problem were systematically presented by the
action group as a problem for both the indigenous population and
the immigrants. The effective anti-heroin campaign and the success
of urban renewal did much to ensure that the action group remained
a significant reference group for the Dutch residents and at the same
time enabled the action group to organize projects in cooperation
with immigrants without being seen as an organization which was
for the immigrants and against the indigenous residents (or vice
versa), a situation which has often been reported in studies of interethnic relations in residential areas (Bovenkerk *et al.*, 1985).

Although the action group's members had an open mind about the further development of interrelations and had proved this by taking the above-mentioned steps, most attention remained directed to the actual process of urban renewal and the control of the heroin problem. In this process, the Dutch residents remained the reference group for the professional welfare workers and the active volunteers in the action group, because the Dutch residents both attended the meetings and participated in any action. Although it is important, for the growth of mutual tolerance, that a norm of mutual tolerance be propagated, it is essential that immigrants should be actively involved in activities and decisionmaking, on a basis of equality. This was the central concern of the Educational Advice Centre. In this group's experience, involving immigrants on an individual basis in decision-making processes and activities in the area did not work. A new model was thus developed: immigrants would be organized into *working parties by ethnic origin*, each headed by immigrants living or working in the area. In these groups, the particular problems encountered by immigrants, which they shared with other residents, should be discussed. Recreational and social facilities should be another important concern. In other words, participants should not only find their participation worthwhile from the point of view of achievement, but they should also enjoy it. Group formation along these lines was seen as the primary means of allowing immigrant residents of the area to participate in the activities and decision-making on a basis of equality (and so to further encourage ethnic tolerance). Through the activities of a young Turkish resident, who first took part as a volunteer and was later appointed as a professional welfare worker, and who succeeded in gaining the confidence of the Turkish residents, a very active Turkish working group came into existence, based on the above principles. The meetings were well attended, and provided an opportunity both for people to meet one another and to discuss problems. Sometimes problems specific to the Turkish communities were discussed, sometimes the emphasis was on common problems, such as education, rent increases and the new construction project. The success of the Turkish group was such that Moroccan and Cape Verdean groups were set up in 1984. The Turkish group took two initiatives which promoted an atmosphere of ethnic tolerance. It provided the impetus for the organization of bi-monthly international Sunday afternoon gatherings, at which residents of all ethnic origins could meet in a pleasant atmosphere. These Sunday afternoon gatherings have always been well attended. It was also due to the efforts of this group that, from 1983 onwards, the action group meetings became bi-lingual: Turkish and Dutch.

The result of all this has been that, in the district after 1980, there was no longer any question of Turks and Moroccans withdrawing into their own circles, as other researchers have found elsewhere (Bovenkerk *et al.*, 1985, p. 29 et passim). On the contrary, the involvement of many of them with the new dwellings, facilities and activities in the district increased considerably after 1980. This is another reason why, for example, full use is now being made of the facilities of the district, and the population distribution in the newly constructed dwellings now reflects the percentages of indigenous and immigrants groups. This is in contrast to many other areas of urban renewal in Rotterdam, where Turks and Moroccans particularly lag behind in moving into newly constructed dwellings (Van Praag, 1986, p. 23). What did this actually mean for the *intergroup relations* in the district?

In a residential situation where many ethnic groups live side by side, being different from other people will rapidly yield a sense of social identity. This applies not only to Turks, Moroccans, etc., but to the indigenous population as well. This sense of social identity will be reinforced by ethnic categorization in the environment. This may be carried out not only by other ethnic groups. but also from within one's own ethnic group. Does this mean that group formation based on ethnic origin, encourages antagonistic images of other groups to be built up and vice versa, and that such group formation will evoke ethnic intolerance? This might easily happen, but whether it happens depends, in my opinion, on the situation in which the groups find themselves, the groups' relations with other groups, formulation of purpose and processes of leadership. As such, conditions in the Old West were favorable. After all, urban renewal took place at such a rate that anyone who wished to do so, could either move into newly built dwellings or soon qualify for these. In this process, immigrants were not discriminated against because of objective housing allocation rules. Moreover, nearly all organizations and institutions took it for granted that the district had an international population. Furthermore, the leaders of the various groups emphasized, besides problems specific to their own group, mutual problems requiring attention, and these leaders were seen as important in district organization. In conditions of the kind described above, the proposal by the Turkish working group to make meetings of the action group bi-lingual, so that they could participate in the decision-making process, was not regarded as a threat, but inspired respect, and the Dutch residents proposed that they should cooperate "because they needed one another to lend force to their campaigns". The very fact that they presented themselves as a group wishing to take part in common activities gave rise to a seemingly paradoxical gain in recognition and acceptance, something they could not have managed individually (Augustin & Berger, 1984). Group formation may be a threat to other groups, but it can also inspire respect and enable its members to attain a position of equality that they had not previously enjoyed. The fact that group formation based on ethnic origin need not necessarily lead to 'ingroup favoritism' and 'outgroup dislike' is, in my opinion, an important point for Dutch interethnic relations (Turner, 1978, pp. 248-250; Deschamps, 1978, p. 155). After all, in the near future at least, group formation based on ethnic origin (especially in the Dutch situation where differences in language and culture between immigrants and the indigenous population are usually large) would still seem to be a prerequisite for the involvement of immigrants in activities and processes of decision-making in situations of housing and work, on a basis of equality (Elwert, 1982, pp. 717-731).

CONCLUSIONS

Whereas, in the Old West prior to 1980, the indigenous residents had been giving the incoming immigrants an unfriendly, if not intolerant, reception, after 1980 a situation of ethnic tolerance developed. Does all this mean that there are no longer any prejudices in the Old West, that people are always friendly with one another and interact as much with the members of other ethnic groups as with those of their own group? There are definite indications that this is not the case. In this study, the dependent variable was not 'liking', but ethnic tolerance: a behavioral concept. I am of the opinion that, in residential areas with a large proportion of immigrants, where there are many different cultural backgrounds and various languages are spoken, as is the case in The Netherlands in a number of old urban districts, what the indigenous population and the immigrants think of and feel about each other is not of primary importance, but how they act towards one another. If interethnic (housing) situations are analysed in terms of behavior towards one another, the concept of discrimination is important. I have described non-discriminating behavior as tolerance. An important advantage of the use of a behavioral concept over the use of an attitude is the fact that behavior and changes in behavior can be observed instantaneously. It is therefore easier to trace effects in that area than in the area of cognitive processes. In addition, an approach which puts behavior and changes in behavior first has an advantage over studies directed at attitudes and changes of attitudes, in that it is easier to indicate, in a process analysis of social phenomena, how interactions can be influenced. Thus, I have tried to show, in this chapter, that ethnic tolerance in this residential area resulted from an interplay of factors. We have observed that both urban renewal and the anti-heroin campaign were fundamental factors in bringing about ethnic tolerance in the Old West. In both cases, many Dutch residents originally defined the situation in terms of opposition between their own community and the immigrant community. However, by defining the situation in the group process in such a way that common interests were put first, and by opting for an approach derived from this, which also did justice to the problems of the Dutch residents, a solution was found that led to a decrease rather than an increase in the conflict.

Also with regard to the intergroup situation, one could initially speak of ingroup favoritism and outgroup dislike. We have seen that, for a long time, the Dutch saw immigrants as competitors and felt threatened by them, and were, therefore, not willing to offer them the same positions, rights and opportunities they offered to members of their own group. Not until a number of basic conditions had been met, could the first steps be taken to improve intergroup relations. As well, in these intergroup processes, a number of factors had to be taken into consideration by the various groups. On the one hand, there was a need for a group formation that would not lead to conflict; on the other hand there was a need for unity without ignoring differences. This was possible because, through leadership processes, people were motivated and situations were defined so that both the individuality and the collectiveness were emphasized. The distinction between these groups was relevant to people. It was thus possible to motivate people to be ethnically tolerant through reference groups, reference persons, and normatively conforming to group norms. By propagating, as a group, the norm of ethnic tolerance, it was possible to make the most of the fact that people already regarded tolerance as an important value and that, since people base their behavior on what other people are likely to think of it, those with ethnic prejudices were marginalized within the group. Moreover, this created the possibility for ethnic tolerance to develop from a passive acceptance of a norm to a more active conviction.

# 9

# ETHNIC STEREOTYPES AND POLICE LAW ENFORCEMENT PRACTICES

Wim Bernasco
Leiden University
The Netherlands

Els C.M. van Schie
University of Amsterdam
The Netherlands

Discrimination against minority groups is a major social issue in most countries and its consequences are felt throughout society. In fact, many discussions have focussed on discriminatory effects in contexts where justice and equity are the most valued norms: police departments, prosecuters' offices and courts. The law enforcement system has been accused of discriminating against less powerful social groups, including ethnic minority groups. Such discrimination need not be deliberate, but may be a natural consequence of bias in the way humans process information, in particular their tendency to rely on stereotypes.

In this chapter we will link several relevant theoretical concepts of stereotyping to empirical findings concerning law enforcement practices of the police. Experimental research has provided valuable insights into the nature and content of stereotypes and the way in which stereotypes influence social perception. We do not intend to delve deeply into theoretical issues, but will focus on the question: Can empirical results with regard to police practices be understood in terms of stereotyping and social attribution theories? We will start this chapter by discussing some relevant concepts in theories of stereotyping. Next, we will consider the question of how stereotyping might influence police law enforcement practices. Finally, we will discuss empirical findings in this field, including results from our own research.

## STEREOTYPES AND ATTRIBUTION

Hamilton and Trolier (1986) define a stereotype as "a cognitive structure that contains the perceiver's knowledge, beliefs and expectancies about some human group" (p. 133). This definition is in line with those of many authors who view a stereotype as a cognitive schema. In the present chapter, we will limit the use of the term stereotype to those beliefs that are shared by large numbers of people. Thus, we will focus on what Ashmore and Del Boca (1981) call 'cultural stereotypes': stereotypes about the content of which general agreement exists within a social group. Although numerous human groups exist which are being stereotyped, during the past decades most research has studied two types: ethnic stereotypes and sex stereotypes. In this chapter, we will concentrate on stereotypes of members of ethnic categories, including both ingroup and out-groups (auto- and heterostereotypes respectively).

The cognitive functions of stereotypes are part of the general cognitive process of perceiving and explaining social behavior. In general, stereotypes help the perceiver to categorize and therefore simplify complex situations by overgeneralizing group characteristics to individual group members. Experimental research has shown that stereotypes have an overwhelming influence on the perception of people. Thus, in a study by Duncan (1976) white subjects showed a tendency to label a certain behavior of a black protagonist as violent, while the same behavior was perceived as playful when exhibited by a white protagonist. In general, information which corresponds to existing stereotypes is more likely to receive attention than information which contradicts the stereotype or is neutral in its content.

Stereotypes influence the explanation of behavior as well. Experimental research has shown that people tend to attribute stereotype-consistent information to stable and dispositional characteristics of the actor, while inconsistent information is apt to be attributed to situational factors. Thus, referring again to the above mentioned study by Duncan (1976), white subjects ascribed violent behavior by black protagonists to personal characteristics, while violent behavior (if labelled as such) by whites was ascribed to situational circumstances. Locksley, Hepburn and Ortiz (1982) note that the explanation of behavior is generally not solely based on the content of stereotypes, but also on other relevant information about the actor. In fact, they find that individuating (non-stereotypic) information is the primary determinant of judgements whenever available, and that stereotypic information affects judgements only in the absence of individuating information. They conclude that stereotype-consistent attributions can be dismissed in the presence of sufficient contradicting information. However, other researchers have questioned this conclusion. Rasinski, Crocker and Hastie (1985), who readdressed the issue, failed to replicate the findings of Locksley *et al.* They found that "(...) subjects seemed to be overcautious in revising their stereotype-based judgments when they were presented with individuating diagnostic behavioral information" (p. 322).

Experimental research into the content of ethnic stereotypes has shown that these stereotypes often contain more negatively than positively valued characteristics. Characteristics like 'aggressive', 'violent' and 'criminal' are considered typical for members of ethnic minority groups (Duncan, 1976, p. 591). It should be noted that, in most countries, these stereotypes are supported by official criminal statistics, demonstrating a higher incidence of officially registered criminal behavior by members of ethnic minorities. These statistics, however, may in turn be seriously biased by selective processing in the law enforcement system. It is evident that stereotypes can become self-fulfilling in this way (Bell & Lang, 1985). This notion is central to labelling theories in criminology, which state that delinquency (or deviant behavior in general) is primarily a function of the tendency to label certain societal subgroups as delinquent and of the consequent autostereotyping by members of those subgroups themselves.

## POLICE ARREST, POLICE DISCRETION, AND ETHNIC STEREOTYPES

A large number of criminological studies, most of them inspired by the labelling approach, have addressed the issue of selectivity in the functioning of social control agencies (Landau, 1978). Offenders from different social groups seem to be treated differently by law enforcement agents (police, public prosecutor and court). Besides sex and social class, the ethnic origin of offenders is often mentioned as a biasing factor. In spite of conflicting evidence, it is expected that, in comparable circumstances, members of ethnic minorities will more likely be stopped by the police (Junger-Tas & Van der Zee-Nefkens, 1977), more likely be arrested once stopped (Hepburn, 1978; Smith *et al.*, 1984), more likely be sent to trial if arrested (Bell & Lang, 1985; Landau & Nathan, 1983), and more likely receive harsher punishment in court (Nickerson *et al.*, 1986). One line of research is concerned with the question of whether ethnic origin is associated with differential treatment by the police in their encounters with citizens. If differential treatment based on the suspects' ethnic origin occurs in police practices, it might play a part in the different stages of the police law enforcement process, which should be clearly distinguished.

In the first stage, police officers have to regard someone as suspect or not, a decision which cannot be based on extensive considerations. Often, unconscious factors will influence these decisions. During patrolling, it will often not be clear at first sight if there are reasonable grounds for suspecting a person of having committed an offence. Thus, where information is ambiguous, stereotypes might prove useful in the decision process. Police officers might show a tendency to be more attentive to behavior exhibited by members of ethnic minorities, a tendency possibly having its origin in the content of the above-mentioned ethnic stereotypes. If so, colored offenders would more likely be stopped for questioning than their white counterparts, just because they are believed to be more likely to commit offences. Selective attention may also be based on victim or bystander evidence, when offenders are identified as members of ethnic minority groups.

In the next stage, police officers are confronted with the decision of whether or not to arrest someone suspected of committing an offence. In this stage, evaluation of the offence and the offender might be expected to influence the decision, besides other factors such as the seriousness of the offence. An offender has a higher probability of being arrested if the offence is attributed to personal characteristics than if it is attributed to the circumstances in which the offence took place. Consequently, as we have explained before, members of ethnic minorities might be more likely to be arrested than citizens who belong to the ethnic majority.

In the last stage, the police function as an (informal) prosecutor, which means that they can either deliver suspects to the public prosecutor or not. This so-called discretionary power is exercised by police forces in all Western countries and is generally considered as practical, reasonable and efficient, especially in the case of juvenile offenders, where the set of alternative dispositions is the largest (Bell & Lang, 1985). In this stage of law enforcement, the explanation of an offence may be influenced significantly by ethnic stereotypes. However, we might expect these stereotypes to influence the first stages of police practice more strongly than the last. In the stages of stopping and arresting suspects, the evaluation of offence and offender seems to be more ambiguous and would necessarily be based on little information, while, in the next stage, police officers can rely on relevant individuating information about suspects and offence (e.g. documentation about prior criminal record, social circumstances, bystander evidence) and dismiss stereotypical information.

Based on the foregoing considerations, we put forward the hypothesis that, as a consequence of stereotyping, members of ethnic minorities are more likely to be stopped and arrested for criminal offences than members of the ethnic majority. Once arrested, they may also have a larger probability of being prosecuted. However, because police officers have more information to rely on in their evaluation of offender and offence, we expect the disadvantages of ethnic minorities to be less distinct at this last stage.

ETHNIC STEREOTYPES AND ARREST DECISIONS: EMPIRICAL FINDINGS

Often, the decisions to stop a potential suspect is based on little, or ambiguous, evidence. There is indeed empirical evidence for the assertion that police officers rely on stereotypes when making arresting decisions. Shoemaker *et al.* (1972) established that positive and negative facial stereotypes exist for certain types of crime. Although their study was confined to photographs of white males, stereotyping based on appearance might be expected to be influenced considerably by the ethnic origin of the rated persons, which is generally a salient aspect of appearance. Indeed, Bodenhausen and Wyer (1985), who requested subjects to make judgements based on a description of a criminal offender, showed that ethnic stereotypes can be activated simply by assigning the target a name which suggests his national or cultural background, without referring to ethnicity in any other way. Although both the study by Bodenhausen and Wyer and the study by Shoemaker *et al.* involved judgements by laypersons, police officers may be expected to rely on stereotypes as well. In a laboratory experiment (Willemse & Meyboom, 1979), slides of faces were shown to police officers and they were asked to rate the persons depicted on 'suspectability'. Subjects relied on stereotypes in their ratings. Men were perceived as more 'suspectable' than women, untidy persons as more so than tidy persons and colored people as more so than whites. This is in line with the results of another laboratory study (Winkel & Koppelaar, 1986), in which police academy students made suspectability judgements of persons interacting with police officers in a video tape film.

In reviewing the field-research literature, Smith *et al.* (1984) state that "while it is generally agreed that blacks are more likely to be arrested than whites, there is no consensus on explanations for this disparity" (p. 236). Observational research (Junger-Tas & Van der Zee-Nefkens, 1977) showed that colored people were stopped for questioning considerably more often than whites. However, the suspicion against them was more often unfounded, suggesting that the stopping of colored people was often based not on actual behavior but on stereotypes. One explanation for this phenomenon (Black & Reiss, 1970; Lundman *et al.*, 1978) focuses on the finding that black complainants request arrest more often than white complainants. Because in the American situation most offender-victim dyads seem to be racially homogenouos, black offenders would be arrested more often than whites for the same offences. Another explanation, suggested by Black (1971) and by Junger-Tas and Van der Zee-Nefkens (1977), states that racial disparities in arrest reflect the more antagonistic behavior of blacks toward police officers. Furthermore, police officers showed more authoritarian and belittling behavior toward colored suspects, while colored people demonstrated more aggressive behavior toward the police than did whites. Thus, hostility may be typical for interaction between police officers and ethnic minorities. Research on minorities' attitudes and feelings toward the police has consistently demonstrated that they have negative attitudes and feelings toward the police (Erez, 1984).

Summarizing the foregoing, we conclude that members of ethnic minorities are more likely to be considered suspect and are more likely to be arrested than members of the ethnic majority. Ethnic stereotypes seem to be at least partially responsible for these disparities, although possibly other factors play a part. The fact that a high percentage of registered offenders are members of ethnic minorities might be due to more frequent police patrols in districts with relatively more colored people. Another explanation for the high incidence of ethnic minorities in the official register might be selective attention and selective report by victims and bystanders.

ETHNIC STEREOTYPES AND THE DECISION TO CHARGE OR CAUTION: EMPIRICAL FINDINGS

A further stage in police law enforcement practice concerns the treatment of suspects once they are arrested. There is an extensive literature on the factors that affect this treatment. This is especially the case with respect to juvenile police departments. In general, police officers can choose either to charge a juvenile, which often leads to formal prosecution, or to only caution him or her, which means that no measures are taken except for registration of the offence in the police records. Much of the literature is concerned with the extent to which legal or semi-legal variables – such as the seriousness of the offence, number of prior police contacts, age of the offender and behavior during the intake interview – actually play a role in the decision to charge, and the degree to which race and other non-legal variables also affect decisions (Bell & Lang, 1985).

Although the research results vary considerably, general agreement exists with regard to the effects of seriousness of offence and prior record. These two variables seem to be the main factors in police decisions. In reviewing previous research, most authors state that the evidence for selective law enforcement mechanisms that would disadvantage members of ethnic minorities remains inconclusive (Smith *et al.*, 1984; Bell & Lang, 1985) or at least questionable (Landau, 1978). This is considered partially due to methodological problems: any study on the effect of race on the treatment of offenders should take into account the impact of other variables that may be confounded with race, which makes it difficult to isolate the effects of individual variables. Thus, if a study establishes racial disparities, one can never be sure that these could not have been explained by some other confounding variables (e.g. social class, suspect's behavior during interrogation or pressure toward prosecution on the part of victims), which were not included in the study.

In our own empirical study, we investigated police decisions regarding juveniles, made during 1985 by the juvenile police department of The Hague (the Netherlands). The data were collected from the official registration documents the police use for record-keeping. The main purpose of the study was to assess whether, in comparable circumstances, the police treat juveniles from ethnic minorities differently than native Dutch juveniles. Juveniles from ethnic minorities included in our study were of Surinam, Turkish and Maroccan origin. Together, these three ethnic groups make up almost half of all offence contacts of the juvenile division of The Hague police department. The documents of 400 delinquents, containing information on 557 encounters with the police, were analysed. It was expected that police officers would be more lenient towards native Dutch offenders so that these would be more often cautioned, while juveniles from ethnic minorities would be more often charged. Stepwise logistic regression analysis (Fienberg, 1977) was used to assess the relative influence of different variables on the police decision to charge. This decision was not influenced by the ethnic origin of the offender. Although the regression analysis showed a small interaction effect of ethnic origin and age, the differences do not confirm the basic hypothesis of harsher treatment of juveniles from ethnic minorities. In accordance with the results of other recent studies, it was concluded that the number of previous police encounters and the type of crime committed are the most important criteria for charging or cautioning. Those who had experienced no, or few, previous police contacts, and those who committed a crime against property or shoplifted, were most likely to be cautioned.

While a large number of studies are limited to statistical analysis of official registrations, only a few studies have taken into account the interaction between suspects and police officers during intake interviews. Extensive studies which included observation and interviews in juvenile police departments (Doob & Chan, 1982; Van der Hoeven, 1986) describe the decision process in the following way: While legal variables like seriousness of offence and prior record are taken into account, police officers primarily try to assess the probability of recidivism. They also evaluate the extent of social control that is being exercised on the juvenile (parental control, general home situation, school integration, neighborhood). Police officers have a tendency to charge if little social control is being exercised on a juvenile. Landau and Nathan (1983) suppose that, in general, less social control is exercised on juveniles from ethnic minorities. This, they believe, is the reason why the use of non-legal criteria in police decision-making disadvantages black juveniles. Furthermore, it is suggested that members of ethnic minorities display more a negatively-valued demeanor in their intake interviews, which is also taken into account by the police. Some researchers further point to the possibility that victims of black offenders are more likely to request formal prosecution of offenders. As we mentioned above, in this case selectivity would be caused by victim behavior rather then police practices. Discrepancies in results might of course also reflect different practices in different police departments.

From our discussion of racial disparities in police treatment of arrested suspects, we conclude that police officers include a large amount of information in their decision making. Their decisions are dominated by legal criteria, while how they use non-legal criteria (such as race) remains inconclusive. If ethnic stereotypes have any effect at all in this stage of law enforcement, they are generally overruled by other diagnostic information, especially offence seriousness and prior criminal record.

CONCLUSION

On the basis of the studies cited, it is not possible to give one clearcut answer to the question: Can empirical results with regard to police practices be understood in terms of stereotyping theories? Results of several laboratory experiments indicate that laypersons, as well as police officers, perceive members of ethnic minorities as more suspectable than whites or natives. Ethnic stereotypes can be activated simply by pictures or names in which the ethnic or cultural background is implied. Such unambiguous results are only found in experiments in which it is possible to manipulate the ethnicity of the target person while other relevant variables are held under control. In field research, it is far more difficult to show any effect of race on the treatment of offenders, because one has to take into account the impact of many variables which are confounded with race. Another feature of laboratory experiments is that, in comparison with real life situations, only very little information is available. If ethnic stereotypes are activated, they will not be overruled by other (diagnostic) information, because little, or none, is available.

As established in field research, the availability of information about the suspect plays a role in the occurrence of racial selectivity in police practices. It has been shown that, in the first stage of police contact when there is relatively little known about the offender, police officers seem to differentiate between races in accordance with ethnic stereotypes. Colored people are perceived as more suspectable and are therefore more often taken to the police station. Racial disparities in police treatment disappear when more information is gathered about the suspect. The police decision is dominated by legal criteria; seriousness of the crime and the number of previous police encounters, while, especially for juveniles, information concerning social circumstances is taken into account. This individuating information overrules the bias based on ethnic stereotypes. However, if a larger proportion of colored delinquents than white delinquents are brought into the police station, this will imply that a colored offender has a higher probability of conviction than a white offender.

In some studies, it has been found that colored people demonstrate more aggressive behavior towards the police, which influences police decisions. However, such antagonistic behavior might be a reaction to the selectivity of the police. The fact that innocent members of ethnic minorities are stopped for questioning by the police considerably more often than whites could be one of the reasons for this uncooperative attitude.

The results of studies performed in the Netherlands are in line with American findings. This is remarkable, because the racial composition of the population differs on many points. In the U.S.A., the population contains a greater diversity of racial background than does that of the Netherlands. Furthermore, the presence of most ethnic minority groups in the Netherlands is relatively recent: Turks, Moroccans and Surinamese have been living there for only a few decades, while the blacks in the U.S.A. have been living there for centuries. It thus appears that, independent of the racial composition of the society, police law enforcement practices are affected by ethnic stereotypes. The disadvantage of minority groups especially appears when little information is available to contradict stereotypes.

# Part III

# Remedies

# 10

# ORGANIZATIONAL INCLUSION OF MINORITY GROUPS: A SOCIAL PSYCHOLOGICAL ANALYSIS*

Thomas F. Pettigrew
University of Amsterdam
Amsterdam, The Netherlands
and University of California
Santa Cruz, USA

Joanne Martin
Graduate School of Business
Stanford University
Stanford, California, USA

American society has entered the second stage of its efforts to eradicate racial discrimination. The first stage, highlighted by the U.S. Supreme Court's 1954 ruling against racially segregated public schools and the Civil Rights Movement of the 1960s, attacked the formal structural barriers to black American inclusion. Many of these formal barriers persist, as demonstrated by the continuing patterns of intense racial discrimination in housing. Yet enough progress has been made in some institutions, notably in education and employment, to introduce a second generation set of less formal obstacles to black inclusion. And these new obstacles more closely resemble the more subtle discriminatory barriers against minorities in such other societies as the Netherlands (e.g., Den Uyl, Choenni & Bovenkerk, 1986).

It has become increasingly clear that the alleviation of formal barriers to inclusion does not in itself ensure black entry into previously all-white American institutions. More subtle barriers have arisen. To be sure, these second-generation barriers are a direct legacy of the nation's three centuries of slavery, legalized segregation, and other forms of legitimized and institutionalized racism. Yet the greater subtlety of these new forms pose new problems of remedy. They act at both the structural, institutional level focused on by sociologists, and the face-to-face situational level focused on by social psychologists. This paper discusses the operation of these obstacles at the social psychological level of analysis.

We advance, then, a social and organizational analysis of the inclusion of black Americans into previously all-white work settings. In particular, we concern ourselves with situations that have often occurred in recent years as a result of affirmative action employment programs. While much attention concerning these programs has been directed on the problems faced by threatened whites, this chapter centers on the situational problems often posed by affirmative action for black employees. We believe that affirmative action in employment is essential for further progress in eroding racial discrimination in the United States and other western nations. But we believe that the ways in which these programs have often been administered have placed an unnecessarily heavy situational burden on those whom the programs are designed to benefit. This chapter analyzes this burden, and illustrates its analysis by proposing potential remedies.

THE PROBLEM

*Gains through governmental action create new interracial situations*

After three-and-a-half centuries of being relegated to America's worst jobs, black workers have been able in recent decades to upgrade their occupations substantially. During the 1960s and 1970s, the black growth in professional, managerial, general white collar, and craft jobs has been more rapid than that of whites (Farley 1977, 1984:48, 1985; Freeman, 1978). This growth began to close the racial differentials in white collar employment. Blacks comprised 10% of the work force in both 1960 and 1980, yet their proportion of those with professional and managerial jobs rose from 3% to 6% in these years (Farley, 1984:194). The 1980 Census found that 24.1% of the entire black labor force was located in the broad white-collar category of 'technical, sales, and administrative support' compared with 29.6% of the white labor force. By 1980, five times as many blacks held these white-collar jobs as those who worked as private household servants -a major black job category in 1960. Although there are limitations to this successful record to be discussed later, these achievements are considerable.

This racial progress in combatting job discrimination over the past generation depended heavily on direct governmental intervention. In helping to achieve a more racially diverse occupational structure, these antidiscrimination programs in employment have had to rely primarily on formal, objectively verifiable means of achieving change, including explicit guidelines and objective criteria for measuring outcomes. But formal, objectively verifiable change programs cannot prevent subjective and interpersonal elements from creeping into organizational life. During the recruitment process, it can be required that minority applicants be interviewed, but the content of that interaction and the recruiter's judgment processes are more difficult to control. Formal training for blacks can be improved, but much of the entry process involves informal learning from peers and supervisors in the course of routine social interaction. Formal performance appraisals can be made more objective, but the process of informal evaluation is continuous. Opinions are often formed and communicated during informal interactions that precede and supplement formal performance appraisals. These areas of working life, where subjective and interpersonal factors inevitably enter, are the vulnerable points of affirmative action programs. Formal, objectively verifiable rules and procedures may restrain but cannot prevent prejudice and discrimination in these situations.

*Modern racial prejudice is more subtle and indirect*

The current extent of American prejudice today is often underestimated, because many people think of white resistance to racial and ethnic change only in terms of raw, overt bigotry. This is misleading. To be sure, old-style bigotry, "dominative racism" in Kovel's (1970) terms, is still often involved. But majority-group resistance to racial change today is generally more subtle, indirect, and ostensibly nonracial. The Civil Rights Movement in the 1960s gave traditional, overt antiblack prejudice a disreputable image.

Modern forms are typified by several characteristics (Pettigrew, 1989). (1) They reject gross and global stereotypes (e.g., blacks as genetically less intelligent) and blatant forms of racial discrimination (e.g., open refusal to hire qualified blacks on racial grounds alone) (Schuman, Steeh & Bobo, 1985; Smith & Sheatsley, 1984; Taylor, Sheatsley & Greeley, 1978). (2) White opposition to racial change is now typically cloaked in ostensibly nonracial concerns. Thus, school desegregation is not attacked directly, but the necessary student transportation to achieve desegregation is singled out for attack (McConahay, 1982; Schuman *et al.* 1985). (3) White American attitudes about racial policy in general, and affirmative action programs in particular, are deeply intertwined with widespread individualistic conceptions of opportunity in America (Kluegel & Smith, 1986).

Schuman *et al.* (1985) document these changes in their comprehensive review of American survey data on racial attitudes gathered over the past 40 years. They note major shifts in white stereotypes of blacks and in matters of abstract principle. In 1942 42% of surveyed whites believed that blacks had the same intelligence as whites; but by 1956 the percentage had risen to 80%, a level where it has remained. More striking are the large majorities of whites who now favor equal racial treatment in a variety of settings. Only 42% of whites surveyed in 1942 thought blacks "should have as good a chance as white people to get any kind of job," but 97% did so by 1972. In addition, over 80% of white Americans now say they would also support having "the same schools" for black and white children, the same public transportation and accommodations for the races, and a "qualified" black candidate of their own political party for President. [1]

The annual rates of change on these questions were greatest between 1963 and 1972, especially the two-year period 1970-72 (Smith & Sheatsley, 1984; D.G. Taylor *et al.* 1978). These were the years of the most sweeping institutional changes in American society. It is this intimate connection between structural and individual changes that is emphasized throughout our analysis. Note also that these questions are broad, hypothetical items that avert personal involvement and threat. The questions do not specify the precise schools or the precise jobs the races will share. Yet the hypothetical quality of these often used survey questions does not render these data meaningless. These dramatic trends are important indicators of a major alteration in white American thinking about the abstract principles of black-white relations. The point is not to overinterpret these data, for they can be – and have been – misread to mean that antiblack prejudice has eroded to the point of near-disappearance over the past generation.

The reality of modern antiblack prejudice is revealed by current white views concerning actual governmental implementation of non-discrimination. Relatively few white Americans favor governmental intervention to secure black entry into jobs (36% in 1974), open housing (46% in 1983), and public schools (25% in 1978) (Schuman *et al.*, 1985:88-90). [2] It is this gap between high principle and low inclination to implement these principles that is the hallmark of modern white American thinking about their black fellow citizens (Pettigrew, 1979b).

The behavioral counterparts of this apparent discrepancy in the current racial attitudes of whites have been uncovered in a variety of social psychological experiments. In a broad range of situations, most whites and blacks (though not all) comply with the new racial norms without full internalization. Often, too, American whites evince either micro-aggressions against or avoidance of black people. These behavioral trends emerge from such studies as Wispe and Freshley's (1971; Gaertner, 1976:198) field study at a Kansas supermarket. Black and white women, matched for social class appearance and age, dropped their groceries as they left the supermarket in the path of oncoming white customers. The dependent variable was how much help the women received. Old-style racist norms would dictate that whites help the white woman but ignore the black woman. This did not happen. In compliance with the new norms, approximately the same proportion of whites stopped to help each woman. But 'the new pattern' is revealed by the unequal help they provided after stopping. On 63% of the occasions with the white woman, the white passersby gave complete help by picking up all the dropped groceries. But 70% of the time with the black woman they gave only perfunctory help by picking up just a few packages – suggesting a failure to internalize the new norms. Forty-three such helping studies reveal varying degrees of discrimination (Crosby, Bromley & Saxe, 1980). Nineteen (44%) noted that subjects gave more aid to their own race than to the opposite race. Black subjects were as likely to demonstrate this tendency as white subjects. Consistent with compliance but not internalization, whites were less likely to discriminate in face-to-face than in non-contact situations.

Nonverbal research, with its focus on behaviors that lie largely beyond awareness, offers additional evidence. These studies often find that white American college students sit further away, use a less friendly voice tone, make less eye contact and more speech errors, and terminate the interview faster when interacting with a black than a white (Hendricks & Bootzin, 1976; Weitz, 1972; Word, Zanna & Cooper, 1974).

*Effects of modern prejudice in the new interracial work situations*

Precisely because of their subtlety and indirectness, these newer forms of prejudice and avoidance are hard to eradicate. Often the black is the only person in a position to draw the conclusion that prejudice is operating in the work situation. Whites usually observe only a subset of the incidents, any one of which can be explained away by a nonracial account. Consequently, many whites remain unconvinced of the reality of subtle prejudice and discrimination, and come to think of their black co-workers as 'overly sensitive' to the issue. Hence, the modern forms of prejudice typically remain invisible even to its perpetrators.

The potential for misunderstanding created by these white views and behavior can be heightened by black reactions. When whites exhibit modern forms of prejudice and fail to see the prejudice inherent in their actions, blacks may respond with anger, alienation, low morale, and other mental and physical symptoms of stress. Not surprisingly, black performance can suffer as a result. When no decrements in black performance occur, this can be erroneously attributed by whites to a lack of discrimination rather than to the extraordinary skills and tenacity of the blacks themselves. This common difference in causal attributions between blacks and whites helps to explain why blacks in these situations firmly believe they must outperform whites to keep their positions while whites often regard their black colleagues as performing at only an average or even below average level (Davis & Watson, 1982; Yarkin, Town & Wallston, 1982).

Confronted with these problems or anticipating their occurrence, blacks will often not apply for, not accept, or not keep jobs in fields and organizations with small minority representation. Because of such patterns of minority reaction, modern prejudice has been termed 'aversive' (Kovel, 1970): blacks turn away from those institutions where their participation is most sorely needed, setting in motion a cycle that defeats the goals of affirmative action.

Grim as this analysis is, social psychological, sociological, and organizational research can help alleviate the effects of modern prejudice. The word 'alleviate' is chosen with care, as the racism of American society makes it impossible for any institution to eliminate these problems entirely. This chapter focuses, then, on the difficulties that are often associated with the inclusion of black Americans in predominantly white employment contexts. The next section of the chapter illustrates some of the ways modern prejudice interacts with situational constraints to create racial discrimination. We discuss three stages in the minority inclusion process: recruitment, entry, and evaluation. The final section describes various interpersonal and structural interventions that can be used to combat these second-generation forms of discrimination.

A SOCIAL PSYCHOLOGICAL ANALYSIS OF THE PROBLEM

*Black Ambivalence During Recruitment*

Under the scrutiny of affirmative action programs, blacks have been encouraged to enter many white-dominated jobs, occupations, and fields of study. More often than advocates of affirmative action would wish, these opportunities for advancement have been rejected by blacks (Heilman, 1979; U.S. Bureau of Labor Statistics, 1982). After a long history of exclusion, it is not surprising that many black people are skeptical when they hear of new opportunities suddenly opening to them. They may assume that their chances of being accepted are low and thus avoid rejection. Even if it is clear their chances of being accepted are high, they may still decide not to apply because they personally or vicariously know the performance difficulties and interpersonal stresses associated with being a minority in a majority-dominated context. This is not another victim-blaming explanation; it is instead yet another cost of America's conflict-ridden racial history that now makes remedial action more difficult.

In part, black ambivalence during the recruitment process may be an aspiration-level phenomenon. Just as aspiration level theory would predict, affirmative action programs can raise black aspirations by increasing the subjective probability of success (Davis & Watson, 1982). In recent years it has become possible for blacks to enter occupations that had been traditionally the exclusive domains of whites. But attacks on the programs that made such entry possible, particularly when widely publicized, fulfill the worst expectations of many blacks.

Ironically, black reluctance may increase when it appears that acceptance into an organization is likely. As long as there is scant chance of acceptance, blacks need not concern themselves with the stresses they would endure were they allowed to enter. But once the opportunity is available, such concerns can cause an unwillingness of blacks to apply. While there are important differences between racial and gender discrimination, research on the effects of sex discrimination on mobility aspirations illustrates the point. [3]

Research on occupational sex segregation has uncovered evidence of women being reluctant to enter jobs previously dominated by males (e.g., Gutek, Larwood & Stromberg, 1985). This refusal of upward mobility has been attributed to several factors: the lack of role models (Kram, 1983); the fear that male colleagues will attribute the hiring of females to preferential treatment and assume them to be incompetent (Heilman & Herlihy, 1984); competing demands for time and energy from husbands, children, and housework (Bernard, 1981; Pleck, 1977); and reluctance to suffer the stress, work difficulties, and prospect of unfair evaluations due to being female in a male-dominated work context (Crocker & McGraw, 1982; Kanter, 1977). Each of these explanations assumes that if these obstacles were removed women and minorities would want upward mobility.

The ambivalence associated with upward mobility may be pychologically very deep. Perhaps in defense against these obstacles to advancement as well as in support of their self-worth, many women and minorities deny the extent to which they are victims of discrimination practices. When the salaries of matched samples of males and females with full-time jobs were compared in one study, the women earned much less despite similar age, education, work experience, hours worked per week, etc. (Crosby, 1982). When questioned, these women willingly agreed that women in general were discriminated against, but they did not see themselves as victims of discrimination.

One explanation for these phenomena may lie in the 'awakening' of depressed aspiration, as suggested by two studies of female secretaries at two large corporations (Martin, Price, Bies & Powers, in press). In Study I, secretarial subjects watched a slide and tape presentation that described a secretarial job and a sales executive job at a company much like their own. The subjects were randomly assigned to one of two conditions. In the first condition, sex segregation was complete: all secretaries were female and all executives were male. In the second condition, sex segregation was reduced: all secretaries were once again female, but a third of the executives were also female.

It was expected that the presence of females in the executive ranks would raise the aspirations of the female secretarial subjects, making them more discontent with their secretarial status and pay and more desirous of promotion. Precisely the opposite occurred. When sex segregation was reduced, the female secretaries were significantly more satisfied with their secretarial status and pay levels, and less likely to want to apply for a promotion to an executive position, even though they thought it likely their application would be accepted. These results were so surprising that Study I was replicated with secretarial subjects from a second and different type of corporation. The pattern of results was the same with even stronger effects. These findings are reminiscent of the approach-avoidance goal gradient phenomenon in motivational psychology (Brown, 1948). At first glance, these results seem to suggest that these female secretaries were glibly saying: "Now that I can have it, I don't want it." But it is more likely, given the cited research on sex discrimination, that there are several quite pragmatic and serious reasons why such a response to opportunities for upward mobility might occur. Objects, such as jobs, are often devalued once they are in plentiful supply (Worchel, Lee & Adewole, 1975). In addition, there is evidence that jobs lose prestige when opened to women or blacks (Touhey, 1974). Post-experimental interviews with the secretaries suggested a third explanation. Excluded groups often do not seriously consider the negative consequences of upward mobility unless that mobility becomes accessible. Then the interpersonal stresses and genuine performance difficulties associated with being a minority in a majority setting become salient and make upward mobility seem less desirable.

These difficulties are heightened when black applicants meet recruiters, especially if the recruiters are not themselves blacks. Consider the results of two experiments by Word *et al.* (1974). In the first study, white Princeton students held interviews with black and white job applicants. Unknown to the interviewers, the applicants were trained confederates who had rehearsed the same responses to all the interview questions. Hence, there were no objective differences in the performances of the black and white applicants. But there were major differences in the behavior of the white interviewers as a function of whether they were questioning a black or white applicant. With a black applicant, there were significantly more low-immediacy behaviors (Mehrabian, 1968). Black applicants received less eye contact, less forward body lean, and shorter interviews – indications of negative interaction. The black applicants also faced interviewers who sat further away and made many more speech errors.

But do these differential responses influence the applicants' performance? To find out, Word and his colleagues reversed their experiment, and used white confederates as interviewers and white subjects as job applicants. The trained interviewers responded to half the applicants with the low-immediacy behaviors that black applicants had received in the first study. The other half received the high-immediacy behaviors that white applicants had experienced earlier. The results are dramatic. The subjects exposed to the low-immediacy behaviors were aware they had been treated coldly; they rated their interviewers as less adequate and less friendly. And when judges later rated videotapes of the second study's applicants, those whites who had been treated 'as blacks' were judged to have been more nervous and to have performed less effectively. The interviewers' behaviors had caused a genuine decrement in applicant performance.

These two interview experiments capture important elements of modern racial behavior. Low-immediacy behaviors are subtle – particularly in contrast to blatant bigotry. Not surprisingly, whites are generally unaware of these shifts in their behavior; typically, they perceive black responses to them as caused not by their own behavior but by something distinctive about blacks. In short, the fundamental attribution fallacy operates (Ross, 1977), and dispositional attributions are more salient than situational ones. Often what is assumed to be distinctive is based on the cultural stereotypes of blacks that currently prevail. Given the historic exclusion of blacks from these jobs, these stereotypes frequently include assumptions of incompetence for critical skills needed in the job situation – punctuality, mathematical ability etc.

For their part, many black Americans become keenly aware of the differential treatment they receive from whites, sometimes so aware they perceive it even when it is not present. Given these difficulties, it is no wonder that blacks frequently react to the recruitment process with ambivalence, as if they were saying: "I'm not sure I'll get accepted; and even if I do, I'm not sure I want the position." Thus, the recruitment process itself can intensify black desires to avoid rejection and interpersonal stress, creating a reduced number of black applications and acceptances. And it is hard to break this pattern, because it is based on avoiding stress and rejection. Avoidance learning reduces the possibility of experiencing corrective situations, such as acceptance and positive interaction. Moreover, experiences as a victim of prejudice and discrimination are highly emotional; and emotional conditioning has a slow extinction curve (Soloman, 1964). Hence, negative black responses to recruitment efforts are especially resistant to change.

*Exaggerated expectations and extreme evaluations during entry*

When blacks enter an organization where they are thinly represented, other problems arise. In addition to the familiar negative effects of stereotypes, recent social psychological research has uncovered detrimental effects created by the new employment situation. Polarized expectations and evaluations, either far too low or far too high, often result from the solo and token statuses thrust upon the new black employees. Because these terms – solo and token – have been used differently by various writers, definitions are needed.

A solo is a single black individual in a group of whites – in Kanter's (1977) terms, an "X" in a field of "O's". More loosely, 'solo' is used to refer to more than a single individual when there are relatively few blacks in proportion to whites in a given work group. Note that the 'solo' concept carries no implications about the reasons why the solo was brought into the group. By contrast, 'token' is used to indicate explicitly that the individual was included in part because of affirmative action considerations. [4] Note that single entrants under affirmative action programs may often hold both solo and token roles. But, when affirmative action efforts effectively generate numerous blacks, they may still be perceived as tokens even though they are not in solo roles.

Usually only a few blacks enter an American organization at one time, with many working groups (especially in the upper organizational ranks) including only one black. Sometimes blacks are put in solo positions simply because they are so rare. Employers may deliberately scatter blacks, one per work group, so as to 'share the wealth' and make the black recruitment efforts highly visible. At any rate, blacks are often placed in solo positions. Moreover, they are often assumed to be tokens as well. This is doubly unfortunate, for solo and token status each can have dramatically negative effects on performance. These effects are stronger during the crucial entry period, when blacks are still strangers and have not yet had an opportunity to prove themselves as competent, 'individuated' human beings.

The first problem most solos encounter is low expectations. White workers often expect blacks to be poor performers, and these expectations may be even lower if the new worker is also a solo. In the words of one insulted solo: "They were astonished to find that I could write a basic memo. Even the completion of an easy task brought surprised compliments." Beyond insulting, low expectations can have a detrimental effect on performance (Rosenthal & Rosnow, 1969). Sometimes this happens because blacks internalize the low expectations of others and come to expect less of themselves. Sometimes whites act on their low expectations in such ways as giving blacks less difficult assignments. Unfortunately, new employees given unchallenging first assignments have lower short- and long-term performance levels than employees given more challenging assignments (Berlew & Hall, 1971). Low expectations, then, can have powerful negative effects on black performance.

Of course, many solo role black workers may, through extraordinary ability and tenacity, prevent low expectations from having a negative impact. Even so, they may receive unfairly low performance evaluations. This point is demonstrated in two laboratory experiments using the same basic design (Taylor et al., 1977). College students, most of them white, experienced a slide and tape presentation that portrayed six teachers discussing new ideas for a school project. The content of the spoken conversation heard by the subjects was the same in all experimental conditions. The race (Study I) or sex (Study II) of the teachers speaking the six roles in the dialogue was varied. Hence, the role of a creative, humorous teacher could be played by a solo black in a group of white teachers, by a solo male in a group of female teachers, or by a black in a balanced group of three whites and three blacks. Following the presentation, the subjects evaluated the teachers' personalities and their contributions to the discussion. Since the content of the contributions was held constant, only race (in Study I) or sex (in Study II) could account for any differences in the subjects' ratings of the teachers.

In contrast to non-solos saying the same words, solos (whether black, female, or male) were rated as talking more, making a stronger impression, having a stronger personality, and being more confident, assertive, and individualistic. Solos, then, were more prominent than non-solos, but this did not give them an advantage. When a contribution was made by a solo, it tended to be rated as less creative and effective than when the same contribution was made by a non-solo. Thus, solo status can negatively influence evaluations even in the absence of any performance problems. Such findings are consistent with the results of a survey of managers from 12 large corporations (Fernandez, 1981). Eighty-three percent of the black managers, in contrast to only 10% of the white male managers, agreed with the proposition that minorities have to be better performers than whites.

Not all solo status effects are unequivocally negative. Rather than greeting the solo with low expectations, a few staunch supporters may have very high expectations for the solo. Unfortunately, these high expectations may be unrealistically high. Similarly, a few performance evaluations may be positively, rather than negatively, biased. Opinions of female solos, for example, often polarize: in the words of an old nursery rhyme, "if they are good, they are very, very good; and if they are bad, they are awful". In the second study by Taylor *et al.*, one of the roles received favorable ratings. When a female solo played this role, her ratings were significantly more positive than when the role was played by a non-solo. A second role was usually considered highly disagreeable. When a female solo played that role, ratings were even more negative.

These results suggest that solos, especially during the initial stages of a working relationship, are subject to exaggerated expectations and extreme evaluations. Either a solo can do no wrong, or else the solo is hopelessly doomed to failure. This polarization of solo evaluations has both advantages and disadvantages. As long as the superior perception can be sustained, the solo offers a counter-example to pernicious stereotypes. The key is maintaining that perception. The obvious danger is that, sooner or later, the solo's performance will regress toward the mean. Then even fully adequate, average work can be a disappointment – particularly to the solo's staunchest allies.

Polarization also raises some less obvious dangers. Faced with unrealistically high or low expectations and evaluations, solos cannot afford to trust the validity of the feedback they receive. Nor are there similar others in the situation with whom to compare. How, then, can solos make an accurate assessment of their performance levels, particularly in an unfamiliar environment? One danger is that solos can become too easily discouraged. Or, defensively, solos may too easily deny the validity of negative feedback, thus overestimating their performance (Larwood, 1982). Either way, the solo is operating without consistently accurate feedback from the environment – as if a plane were flying without a gyroscope.

In addition to the problems of solo status, blacks are often assumed to be tokens as well. Unfortunately, the attribution of preferential treatment that is the hallmark of token status also brings its own, uniquely negative consequences. The most important is the assumption of incompetence (Heilman & Herlihy, 1984). The 'logic' of this assumption is that new black appointments must be incompetent, since if they were competent they would have been hired without help from affirmative action. Notice that this reasoning requires the denial of racial discrimination in contemporary America – a widespread belief among white Americans today and an important underpinning of modern prejudicial thought (Kluegel & Smith, 1986).

The results of an experiment by Northcraft (1982) are relevant here. White subjects were provided with five resumes representing five individuals who had supposedly been hired as investment coun-selors. According to a panel of judges, three of the resumes had 'good' qualifications, one 'intermediate', and the fifth 'poor'. The subjects were told that one of the resumes belonged to a black male, one to a red-haired white male, and one to a white male who enjoyed golfing. The subjects' task was to guess which of the five resumes belonged to each of these three individuals. The dependent variable of interest was what per cent of the subjects would guess that the poorly qualified resume belonged to the black candidate. If no negative racial expectations influenced subjects' judgments, the poor resume should have been assigned to the black candidate by chance 20% of the time. Instead, 37% of the subjects selected the 'poor' resume as belonging to the black. This significant departure from chance was exacerbated in a second condition. New subjects were given the same task with the additional information that affir-mative action was involved in the hiring process. In this condition, 69% of the subjects guessed that the 'poor' resume belonged to the black candidate. As in the studies of sex discrimination, the mention of affirmative action was assumed by whites to mean that blacks were hired only because of preferential treatment and not merit.

Not only are tokens assumed to be incompetent, they also must bear the usual difficulties associated with negative racial stereotypes. A token is distinguished from peers on two dimensions: race and affir-mative action. Whites often jump to the conclusion that the token must be dissimilar on other dimensions as well. As Allport (1954) noted, if you give people a thimble full of facts, they will fill a bath-tub with inferences and generalizations. [5] When people are seen as dissimilar, they are more likely to be disliked and avoided (Rokeach, 1960; Berscheid & Walster, 1978; Wilder & Allen, 1974). For this rea-son, tokens are often lonely in spite of their best efforts to establish friendly relationships with their colleagues at work. In an investiga-tion of large corporations, a substantial majority of the black man-agers felt that minorities were excluded from formal work groups (Fernandez, 1982). Exclusion is not just a social problem; it can importantly detract from black performance, especially in work set-tings that require cooperation and teamwork.

The most relevant research on this point involves women. Kanter's (1977) token female sales trainees were assumed to be dissimilar on a variety of dimensions not directly related to their competence or gender. They were often not involved in informal conversations or social outings, and soon suffered from severe isolation. As a result, the tokens were less likely to be included in work-related exchanges of information, their morale sank, and negative effects on their performance were soon evident.

The Northcraft and Kanter studies suggest that, in the absence of countervailing information, majority workers will assume that tokens are incompetent and personally dissimilar. Would these preconceptions persist if countervailing evidence were presented? To find out, Northcraft and Martin (1982) provided white students with resumes, including photographs, of advanced students. The subjects' task was to choose one of these advanced students to lead a review session in preparation for examinations. For one of the resumes, changes in the photograph and content were made so that it appeared that this student was either white, black, or a black token as indicated by the label of 'affirmative action student'. In addition, the amount of information about the qualifications of these advanced students was manipulated to be either minimal (departmental affiliation only), clearly well qualified, or well qualified plus personally similar to the subjects. When information about qualifications was minimal, the target resume was selected significantly less frequently when it appeared to belong to the token black than when the same resume appeared to belong to either the white or the unlabeled black. When full competency information was provided, the bias against the token disappeared, although a slight tendency to avoid the other black remained. When both competency and personal similarity information was given, all evidence of bias disappeared. These results suggest an optimistic conclusion – that the biasing effects of token status can be alleviated if information about the competency and personal similarity of the minority member is conspicuously presented.

But an additional investigation took a deeper look at the efficacy of this strategy. Business school subjects, almost all of whom were white, were asked to read a case study about a high pressure consulting firm and to imagine that they were one of four aspiring junior consultants competing for a coveted managerial promotion (Northcraft & Martin, 1982). Subjects were given a set of resumes, including photographs, representing the three other junior consultants. The task was to choose one as a co-worker on a project. The resumes indicated that two of the junior consultants were impressively well qualified, while the third was minimally qualified. For one of the well-qualified resumes, changes in the content and photograph were made so that it appeared that the junior consultant was either white, black, or a black token with the label 'affirmative action appointment'. Personal similarity information was also either present or absent in all the resumes. The target resume was clearly well qualified, and whether it belonged to a white, an unlabeled black, or token black, the subjects selected it. Yet subtle evidence of prejudice emerged. When the target resume belonged to a white, the presence of personal similarity information caused an increase in the ratings of the co-worker's probable competence and the anticipated satisfaction with the working relationship. When the target resume belonged to a solo black, similarity information had no effect. When the target resume belonged to a token, the presence of similarity information caused a decrease in the ratings of the co-worker's probable competence and the anticipated satisfaction with the relationship. Thus, the presence of direct information about competence and personal similarity removed some, but not all, of the negative consequences of token status.

New black employees, then, are in triple jeopardy at the entry stage. First, they must cope with negative stereotypes about their abilities. Second, if their group is thinly represented, they are subject to the largely negative effects of solo status. They may have to cope with polarized feedback – unrealistically low or high expectations and extreme evaluations. Whether deemed wonderful or worthless, solos do not receive the realistic feedback that career development requires. Their job performance can suffer. And even if it does not suffer, evaluations of it may be biased. Tokenism represents a third source of jeopardy. If blacks are labeled as affirmative action appointments, whites may assume they are incompetent and not wish to work with them. Their personal dissimilarities to majority members may be exaggerated. They may be disliked and become socially isolated. These biases can be partly diminished by the provision of explicit competence and personal similarly information, but subtle residues of prejudice may remain.

*Cognitive Distortions During Evaluation*

Because blacks are more likely than others to leave an organization during the recruitment and entry stages, the likelihood of solo status increases as they continue to stay and rise higher within a hierarchy. Moreover, once they pass initial screenings, the threat to established white job holders is more direct. Consequently, the possibilities for biased perceptions and evaluations of black performance should increase at this point in the inclusion process. There is considerable evidence that blacks, particularly if they are high performers, receive unfairly low performance evaluations. In one study the performance levels of black and white workers were manipulated (Hamner, Kim, Baird & Bigoness, 1974). White subjects, who ranked the performance levels of these workers, showed clear evidence of systematic racial differences. Among workers with equally high performance, black workers received significantly lower evaluations than white workers. Further, among low performing workers, blacks were rated more highly than whites. Social psychological research offers an explanation for these evaluation biases, one that suggests that bias occurs primarily for cognitive reasons, although emotional sources of prejudice may shape the specific form these biases take.

Ever since Walter Lippman (1922) coined the term 'stereotypes', it has been understood that our perceptions and evaluations of human groups are shaped by our prior beliefs. We construct categories in order to organize our understanding of the social world. Stereotyped beliefs consist of associations we make between two kinds of categories: types of people and clusters of particular human characteristics. Lippman emphasized that these "pictures in our head" are generally not the product of personal experience. Instead, these images are provided by our culture and reinforced by the mass media. Furthermore, extensive social cognition experimentation shows how stereotypes, acting as social schemata, shape perception and memory as well as evaluation (Fiske & Taylor, 1984:159-167; Hamilton, 1981; Miller, 1982).

Once established, stereotypic beliefs are extremely resistant to change. An array of social psychological studies show the many ways we maintain our beliefs about people in the face of contrary evidence. Biased information seeking (Snyder & Cantor, 1979), labeling (Langer & Abelson, 1974), and recall (Cohen, 1981) processes all preserve our social stereotypes. Hence, veteran white employees may seek out and remember facts about new black personnel that are consistent with their low expectations for them. Later, through encoding and retrieval biases, white employees may be more likely to recall those aspects of black behavior consistent with their initial stereotypes and expectations.

Causal attributions also contribute to belief perseverance. With 'confirmatory attribution' (Kulik, 1983), we attribute causes for a person's behavior in a manner that confirms our beliefs about the person. Taylor and Koivumaki (1976) found that negative behavior is typically attributed to situational factors for intimates ("She isn't really like that at all; she just had a bad day.") but to dispositional factors for strangers ("He certainly isn't a very nice person, is he?"). Intimates, then, receive 'the benefit of the doubt'; strangers do not. We can explain away belief-inconsistent behavior with situational attributions and use belief-consistent behavior to confirm our beliefs with dispositional attributions (Crocker *et al.*, 1983).

Pettigrew (1979a) formalized the application of these ideas to problems of prejudice against members of minority groups. (See also Hewstone's paper in this volume.) He proposed an 'ultimate attribution error', whereby attribution processes support negative stereotypes of minorities. When blacks behave in a manner perceived to be negative, whites, especially those who are prejudiced, are likely to attribute their behavior to the personal character of the blacks. Often these dispositional causes are viewed as unalterable innate attributes. From this perspective, situational determinants of negative black behavior, such as solo status, are ignored.

'The ultimate attribution error' also applies to black behavior that is seen by whites as positive. Whites (again especially if they are prejudiced) can explain away the positive behavior in one of four ways. First, an exceptional black may be contrasted with other blacks, thus making 'the exception prove the rule' (as with the 'pet' solo whose virtues are exaggerated and whose presence 'proves' there is no prejudice or discrimination in the organization). Second, high levels of black performance can be attributed to such transitory causes as luck or unfair special advantage, rather that to the skills and abilities of the black employee. Accusations of unfair advantage are particularly likely if the black person is regarded as a token. Third, top performance by a black can be attributed to high motivation and effort. This is often seen as an unfair violation of the norms, a compensation for lack of talent, and ultimately unsustainable. Finally, high levels of black performance may simply be attributed to situational factors, such as the availability of good equipment or plentiful assistance. These attributional tendencies are intensified in situations where: group identifications are salient; the perceiver is directly affected by the minority group's behavior; social class covaries with minority and majority group membership; and negative stereotypes are widely held (Pettigrew, 1979a). Both solo and token statuses tend to accentuate these conditions.

Relevant research supports the ultimate attribution formulation. For instance, Hindu subjects in India employ dispositional attributions to explain positive acts of Hindus and negative acts of Muslims (Taylor & Jaggi, 1974). Another study examined the causal attributions of white undergraduates for a successful banking career; both female and male subjects attributed significantly less ability, more effort, and more luck to a white female, a black male, or a black female than to an identically successful white male (Yarkin, Town & Wallston, 1982). A detailed examination of the ultimate attribution formulation used a quiz contest format (Pettigrew, Jemmott & Johnson, 1986). White college subjects, serving as observers, watched two male students engage in a contest – one as questioner, displaying his personal store of esoteric knowledge, the other as contestant. Though apparently chosen at random, the contest players were actually confederates enacting a rehearsed scenario of questions and answers. In this situation, questioners have a tremendous advantage in that it appears they have a large store of knowledge much of which is unknown to contestants. In fact, contestants may have equally large or even larger stores of knowledge, but they are unable to display it. Previous research had shown that observers in this situation tend to discount the enormous advantage held by the questioners and rate them as far more knowledgeable than the contestants. Indeed, this effect tends to be strongest for the contestants themselves (Ross, Amabile & Steinmetz, 1977). Pettigrew *et al.* (1986) replicated this 'questioner effect'. They also varied the racial composition of the pairs of contest participants. The white observers tended to explain away both the 'impressive' performances of black questioners and the 'poor' performances of the white contestants by attributing them to transient factors. Consistent with stereotyped beliefs about the greater intellectual competence of whites, they saw the black questioners as lucky and highly motivated and the white contestants as anxious, unlucky, and unmotivated. In contrast, the white questioners' strong performances and the black contestants' weak performances were more often viewed as indicative of true ability and educational attainment. A serendipitous result from this study underscores the force of this effect. Of the initial 116 subjects, 18 expressed suspicion at the close of the experiment that the study had involved racial issues. Overwhelmingly, these suspicious subjects had been assigned to conditions where the 'impressive' questioner was black. As Freud demonstrated, the most primitive means we have of protecting our beliefs is simple denial. These suspicious subjects, faced with evaluating a black who violated their expectations, rendered the inconsistent information meaningless by holding it to be counterfeit.

A follow-up experiment demonstrated the depth of the phenomenon and the difficulty of overcoming it (Johnson, Jemmott & Pettigrew, 1985). Even those subjects who recognized the substantial situational advantage enjoyed by the questioner revealed a strong tendency to view the questioner as superior to the contestant in knowledge, memory, and education. Though less biased than subjects who remained unaware of the situational imbalance, the aware subjects failed to make full use of their valid understanding of the question-and-answer contest when asked to make trait inferences about the questioner and contestant. These results bear dour implications for mere informational remedies. They suggest that efforts to point out the situational determinants of the behavior of solos, tokens, and blacks will by themselves be insufficient to prevent dispositional biases in the evaluations of black performance.

Note that these studies have not focused on highly prejudiced and poorly educated subjects. Participants in the questioner and contestant study were Harvard University students. It is therefore reasonable to presume that the biases of the ultimate attribution error will emerge in many types of organizational contexts, including high ranking and well-educated employees. There is, then, good reason to suspect that formal and informal evaluations of blacks in these organizations will be biased. This makes it difficult, as well as stressful, for black personnel to demonstrate their abilities, even when their performance is equal or superior to that of others.

These evaluation biases raise problems for black retention in two ways. First, failure to receive expected promotions within an organization is a major reason for all personnel to consider leaving. If evaluation biases discriminate against black promotion, then low retention rates will result. Second, these biases are by no means universal in American institutions. But their existence in an organization fulfill black fears, lead to stress, and can lower black performance. Beyond promotional problems such stress contributes to the desire of black workers to leave. Here a labor market point is relevant. Blacks with modest skills are often trapped in discriminatory work situations. But this is not the case for those with specialized skills or experience in high ranking positions. Retention rates for these highly skilled black personnel, compared to other blacks, are far more sensitive to the fairness of evaluations and the severity of stress. Consider the results of a study of turnover in eight major corporations. Highly valued minorities who left management positions cited prejudice and unfair performance evaluations as the causes far more often than comparable majority managers who more often cited little chance of advancement (Jones & Olmedo, 1986).

THE SEARCH FOR EFFECTIVE REMEDIES

*Social Psychological Remedies*

In contrast to the traditional view that racial behavior will change only after racial attitudes change, modern social psychology emphasizes that altered behavior is more often the precursor of altered attitudes (Aronson, 1969; Bem, 1972; Festinger, 1957; Pettigrew, 1961). And behavior is shaped in important ways by the situation in which it occurs. The major remedies we shall mention are based in varying degrees on the premise that shaping behavior by shaping situations is more successful than direct attempts to change such deeply held attitudes as racism.

Some American institutions have developed and refined several techniques for combatting prejudice, including sanctions by high-ranking authorities, training programs, restructuring of task and team assignments, and altered reward systems. Such techniques have had some success in combatting prejudice and discrimination, especially in its overt forms. But it is important to note that there are critical limitations to this record of success. Black occupational upgrading is concentrated in particular sectors of black America.

Gains are strongest among women, the better educated, and younger members of the black labor force (Freeman, 1978). Black increases in upper-level jobs are heavily dependent on public-sector employment (Collins, 1983; Hout, 1984). And convergence with the white occupational distribution has tended to occur in those occupations where white labor force share was already stable or declining. Black shifts out of such low-status occupations as domestic servants and laborers were achieved partly by higher rates of black absence from the labor force (Hauser & Featherman, 1974); that is, part of the decline in low-level black employment came about through blacks simply leaving the labor force rather than securing better jobs. Even the remarkable black entry into white-collar employment dims when placed in historical perspective. By 1982 30% of employed nonwhite men and 52% of employed nonwhite women occupied such positions, but these percentages were reached by white men and women in 1940 (Farley, 1984:50).

Old techniques need revision to counter the more invidious forms of modern prejudice that currently hinder black inclusion. Over three decades ago, Kenneth Clark (1953) held forceful authority support for institutional change to be the key to successful racial integration. Authority sanctions for affirmative action programs are now commonplace. Research on the effects of token status suggests new forms authority sanctions might take. When blacks are hired, especially if they are affirmative action appointees, authorities must make an effort to circulate information about the competency and personal characteristics of all new hires. Instead of providing abstract statements, specific skills and interests should be described in detail. Such information could partially counteract the assumptions of incompetence and personal dissimilarity that impede the inclusion of black workers. Authorities should also give a clear description of affirmative action procedures. It is often wrongly assumed that such programs operate by setting up a separate track for blacks and selecting 'the best' from this smaller group. Such a separate track was specifically outlawed by the U.S. Supreme Court in 1978. It should be made clear that hiring procedures are made in two stages: competence is the first screening criterion, and, then, those minorities who have survived this first critical screen are given some measure of preference.

Training programs should also be used to combat the modern forms of prejudice and discrimination. In one study of corporate managers, both whites and blacks as well as males and females were more likely to feel white males needed special training more than minorities or women (Fernandez, 1982). We concur. Although blacks might benefit from certain specialized skill training, any such training risks blaming blacks for the discriminatory problems they encounter. It is less questionable to consider training white supervisors and high ranking superiors to understand how the structure of an organization, as well as their own behavior, can have discriminatory effects.

While most black managers surveyed by Fernandez (1982) felt that white males needed special training, many did feel that such training would be ineffective. The research cited above, however, suggests how the content and methods of such training might be improved to enhance effectiveness. For example, recruiters could be videotaped as they interviewed black and white applicants. The videotapes could then be analyzed for signs of the low-immediacy behaviors that cause decrements in applicant performance. Similarly, those who do formal performance appraisals could review past evaluations to determine if they personally have a tendency to set less challenging objectives for black subordinates, or if they are more likely to attribute black successes to factors other than ability. The premise of such efforts is that the trainees will alter harmful behavior once they become aware of it, either because they do not want to discriminate or because they fear sanctions from their superiors.

Another approach, suggested by Allport's (1954) intergroup contact hypothesis, is to facilitate racial interaction in task settings by encouraging common goals and intergroup interdependence. One way to create common goals is to structure the situation so as to induce cooperation rather than competition. Many tasks must be performed by a team, working effectively together. Jobs, tasks, and team assignments can be redesigned so as to maximize racial interdependence (Hackman & Oldham, 1976; Sherif, 1966). Information provides its holder informal power. Similar benefits flow from socially or functionally central jobs. Some skills or jobs are central to the success of a team effort, and they should often be strategically assigned to black members. This redesigning effort would give blacks informal sources of power by making the group dependent on their contributions. Research by social psychologists on the application of this principle to interracial classrooms in the form of student team learning (e.g., the jigsaw classroom) has consistently produced beneficial results (Aronson, Stephan, Sikes, Blaney & Snapp, 1978; see also Van Oudenhoven, this book). Such efforts would combat the assumption of incompetence, since contributions of black workers would be apparent to their white co-workers. Whites would also acquire a stake in increasing the skills of their black colleagues. Moreover, during informal interaction among team members, common interests would become more evident and counteract assumptions of dissimilarity.

Interdependent work roles also have the potentiality of having favorable white attitudes toward the black co-worker generalize to broader racial attitudes. In an extended series of experiments, Cook (1984:183) confirms earlier research: "Task interdependence induces cooperative, friendly behavior and develops liking and respect for one's [multi-ethnic] groupmates." But Cook goes further. His studies introduce explicit procedures for ensuring that his white participants understand how the adoption of equalitarian social policies make it less likely their new black friends will encounter discrimination and prejudice. He shows that this individuation procedure promotes "the generalization of positive affect felt for other-ethnic friends to race-relations policies that would benefit them". Thus, using the salient case of a well-liked black colleague as "the foot in the door," Cook achieves generalization to blacks as a group by deliberately making group categorization salient so as to personalize prejudice and discrimination. [7] Field research suggests that this laboratory procedure is not an uncommon occurrence in contemporary organizations (e.g., Kanter, 1977).

Our suggestion concerning interdependent work roles runs counter to the recommendations of those who advocate greater personal independence and responsibility for minority workers together with personally identifiable outcomes (e.g., Fernandez, 1982). These advocates see their recommendation as a means of limiting the racially biased, individual performance appraisals. The dangers associated with such a policy are that black isolation and loneliness could easily be worsened. Although neither unfair appraisals nor isolation is attractive, the trade-off between them is difficult to access. But whether or not interdependent work roles are adopted, it is essential to recognize the possibility, even probability, of unfair performance evaluation with its correlates of unfair pay and restricted promotional opportunities.

Because blacks have to cope with many extra difficulties, the performance levels of some will suffer – particularly those struggling with the triple jeopardy of being a black, solo, and token. If groups and individuals are rewarded strictly by 'merit' or seniority, without consideration of the special burdens shouldered by minorities, blacks then become potential liabilities. Yet if behavior change is desired, it must be rewarded. Otherwise an organization is committing the folly 'of desiring B while rewarding A' (Kerr, 1975). Thus, such potential liabilities can be alleviated by alterations in an organization's reward system. For instance, a small part of the pay raises of supervisors and high-ranking officials could be made contingent on the performance levels of their black subordinates. Similarly, black and white peers could be evaluated in part according to the performance of their work as a team.

Promising as such 'micro-remedies' have proven in certain institutions, their success is heavily dependent on the larger organizational context within which they are applied. Each has definite limitations unless it is accompanied by broader structural changes. Thus, authority sanctions can easily be dismissed as corporate propaganda: "They are just saying that for the record so they can get federal contracts." This problem can be alleviated if noble words are coupled with forceful policy changes and supervision embedded in structural alterations that affect all parts of the organization: "It looks like they really mean it!" Unfortunately, the subtle interpersonal behaviors characteristic of modern prejudice are difficult to detect. Furthermore, unless the organization's fundamental norms and expectations are modified, mere compliance will disappear whenever surveillance is removed.

Similarly, training programs designed to alleviate modern forms of prejudice are extraordinarily difficult to design, deliver, and evaluate. Their underlying assumption – that awareness of personal prejudice will lead to altered racial behavior – is undercut by the highly developed defenses many white Americans have concerning their own prejudices and the justice of the society in which they live and work. Indeed, one of the distinguishing characteristics of modern racial prejudice is that many whites deny personal prejudice even to themselves, and carefully avoid situations where evidence might emerge that conflicts with this self-conception (Kovel, 1970). Considerable resistance to such training programs, then, can be anticipated. And this resistance is likely to be heightened by the focus of the training itself – on subjective and interpersonal issues that may seem far removed from white trainees' job-related concerns. The training content itself might even seem irrelevant to prejudice, if trainees view prejudice as just overt bigotry. In addition, given the difficulty of monitoring the desired behavior change, on-the-job follow-up and evaluation of the training are not easy. Lastly, when turnover or promotion removes trainees from the field of concern, the training must be repeated with their replacements.

Task and team redesign are more likely to have beneficial effects as part of larger organizational alterations. But, if such changes are implemented in isolation, help for the problems discussed earlier is unlikely. Many, if not most, blacks would still be solos and tokens. Many white supervisors would still systematically employ cognitive biases in their evaluations of blacks. Most importantly, the team and task redesign activities would reallocate only informal sources of power among peers. Again, without wider changes, the more formal sources of power, such as line management authority, would remain undisturbed.

Making whites' rewards partly contingent on black performance could create intense resistance and procedural problems. If blacks are successful, how much of their success is due to their superiors? In many cases, black employees may have succeeded despite their supervisors. Likewise, a black's low performance might occur despite a supervisor's best efforts. It would be frustrating for supervisory personnel to have their rewards tied to outcomes they feel they could not control (Porter & Lawler, 1968). It could also be frustrating for black subordinates to have their accomplishments credited to uncooperative superiors.

These micro-remedies are social psychological; they focus on interaction among people currently located within a given setting. These and similar remedies are an attempt to make the best of a bad situation by assisting minority members who are currently coping with the difficulties of being the few among many. This is certainly a worthwhile, if limited, objective; and organizational norms may well be slightly improved. But micro-remedies are costly and time consuming. They are formal, overt attempts to alleviate prejudice that is, by its nature, informal, and covert. More importantly, these micro-remedies accept as a given the current distribution of blacks within organizations: (a) small numbers entering the organization at any one time, (b) scattered across functional areas, (c) so that solos are the norm within any one working group with (d) blacks located overwhelmingly at the lower levels of the hierarchy. Because micro-remedies accept this distribution, they attack only the symptoms rather that the root causes of the problems of black inclusion. Indeed, these narrowly focused efforts may even serve, wittingly or unwittingly, to divert attention away from the fundamental structural issues. Consequently, if micro-remedies are not used in conjunction with broader structural changes that alter the organization's personnel distribution, they are unlikely to have a broad and lasting impact on black inclusion.

*Organizational and Sociological Remedies*

The structural approach of organizational and sociological research suggests methods for alleviating causes, not symptoms, of the modern forms of racial prejudice and discrimination. Strategic alterations in organizational structure have the potential to make the micro-remedies of social psychology effective, while simultaneously easing the problems that these techniques address. A key cause of the problems discussed in this chapter is the small number of blacks in the applicant pool and at the higher levels of organizational hierarchies, where power is exercised, policies formulated, and evaluations made. There are two aspects, then, to this structural analysis: sheer numbers and the lack of power. We shall illustrate our point with several among many potential means to attack both of these issues.

The only conclusive way to prevent the discomforts of solo status and the need for creating tokens is to obtain a critical mass of black representation. Three points about this objective require clarification. First, an accumulation of members of a variety of minority groups, with just a few representatives of each group, does not touch the problem. Such an accumulation has in common only minority status in the eyes of the majority. To address the issues of solo status and tokenism, there must be a critical mass of a given minority group. But what constitutes 'a critical mass'? Two different lines of research – one on the acceptance of women in previously all-male employment situations (Kanter, 1977), the other on racial integration of public schools (Pettigrew, 1975:238) – each led to a similar definition: roughly 20% (but not less than two individuals). [8] In settings such as a school, a minority of 20% represents a significant portion of the student body, large enough to be filtered through the whole school structure. In those fields, labor markets, and areas where the scarcity of qualified minorities is a problem, what can be done to achieve a semblance of a critical mass? We have two suggestions: clusters and tithes.

When blacks are scarce, they can be clustered rather than scattered. That is, black workers could be assigned in small numbers (of not less than two individuals) to groups otherwise composed of whites. This technique reduces the probability of solo status and gives each black a small support group of similar others. Though clustering is not the equivalent to the creation of all-black groups, the obvious danger is that they may nonetheless come to be perceived as a low-status ghetto by both blacks and whites. To avoid this, clustering must be seen as a transitional structural device; and it should be done with work groups fully comparable to all-white work groups in the same organization. In short, care must be taken to avoid any resemblance between these temporary clusters and the traditional patterns of low-status, low-wage, deadend employment ghettos that mark the history of black American employment. Of course, in some cases clustering will be either impossible or inadvisable. For example, the numbers of blacks at the top levels of organizational hierarchies are likely to remain small until more blacks can survive and work their way up the career ladder. In the meantime, it is essential to give as many blacks as possible the formal power that comes from high status, even if it means these individuals will have to suffer the slings and arrows of solo and token status. Together with wielding power, these high-status blacks can have a voice during policy formulation, serve as visible symbols and role models of success, and, if they are willing, be mentors and watchdogs for other black members.

In addition to clustering, organizations could 'tithe' themselves by investing a very small percentage of their personnel budgets in training blacks with precisely those skills needed by the organization. Although tithes refer to a 10% contribution, a tiny fraction of this percentage of a large corporation's personnel budget could have a highly leveraged, targeted effect on a local labor market. This tithe could 'trickle down', most of it being invested in educational programs that have produced current employees, some of it for education in specific, underrepresented subject areas, and a portion going to local schools that meet specified objectives, such as graduating a certain percentage of black students who have completed mathematics and natural science courses. While tithes could be voluntary, the power of the device could obviously be enhanced if it were encouraged by governmental policy.

Clusters and voluntary tithes are relatively low-cost remedies. They can directly affect the supply of qualified blacks, thus shaping an organizational context within which micro-remedies can be implemented more successfully. Fundamentally, macro-remedies have more potential than micro-remedies to render a major, lasting influence on the minority inclusion process. But the principal thrust of our analysis leads to the conclusion that both types of remedies should be used together to make extensive minority hiring and promotion a reality.

A FINAL WORD

From this analysis, one fundamental feature of these newer concerns invites special attention. Current interracial interaction issues appear to have two sources: antiblack prejudice (in both its traditional and modern forms) and typical features of current interracial situations (e.g., the triggering of cognitive biases by solo and token statuses). Each seems to be a sufficient, though not necessary, cause of racially discriminatory behavior. But we do not as yet have an adequate understanding of how these two sources interact, how they each shape the effects of the other. Presumably, both factors are influenced by the larger organizational context within which they are embedded. What has been missing from social psychological analyses of these newer racial problems, with few exceptions (cf., Gaertner, 1976; Gaertner & Dovidio, 1977; Weitz, 1972), is research that examines the joint operation of both prejudice and situational factors in a variety of critical social contexts. Such research offers a useful agenda for the future.

NOTES

* An earlier, more extensive version of this chapter appeared as: T. F. Pettigrew & J. Martin, Shaping the organizational context for black American inclusion, *Journal of Social Issues*, 1987, 43 (1), 41-78.

[1] Racial intermarriage is the only exception to these trends among these items. Only 40% of whites by 1983 "approved" of intermarriage, though even this represents a significant increase from only 4% in 1958 (Schuman *et al.*, 1985:74-75).

[2] These results are not simply a general rejection of governmental intervention into all racial matters, for once again there is an interesting exception among these items. By 1974, two-thirds of whites agreed that the government "should... support the right of black people to go to any hotel or restaurant they can afford" (Schuman *et al.*, 1985:88-90).

[3] We have discussed research on sex discrimination where it is relevant to our argument and where we could not find comparable studies of discrimination against black Americans. We would have liked to have extended parts of our argument to consider 'the double discrimination' faced by black women. Unfortunately, research on this important issue is rare. It is, for example, unclear whether these two forms of discrimination operate independently or, as we suspect, interactively.

[4] These definitions differ from those used by Kanter (1977).

[5] We are indebted to Dr. Greg Northcraft for drawing this Allport quotation to our attention and for contributing many of the ideas that are discussed here on the effects of solo and token status.

[6] Linville and Jones (1980) see this phenomenon as part of a general tendency of outgroup polarization. In a series of experiments with white subjects evaluating law school applications, they found that a black applicant with strong credentials was judged more favorably than an equally strong white applicant. Likewise, they found that an application from a black with weak credentials was rated more negatively than an application from a white with equally weak credentials. Linville and Jones further showed that this polarization effect, similar to that found for solos discussed earlier, was related to the more complex schema their white subjects were able to employ for judging their own group. More complex schema, for instance, allow whites to find more 'excuses' for the poor performance of other whites.

[7] Note this increase in the salience of group categorization is directly the opposite procedure from the exclusive reliance of Brewer and Miller (1984) on low group salience. But the contradiction is more apparent than real (Pettigrew, 1986). Limited salience of group categorization facilitates initial attraction to a cross-group member. But high group salience facilitates the Cook effect – the generalization from liking an individual outgroup member to becoming more

favorable to the outgroup as a whole. Consistent with this resolution, several investigations have found that generalization effects are strongest when optimal contact is experienced with a representative member of the outgroup (Weber & Crocker, 1983; Wilder, 1984).

[8] For small minorities, the implementation of this suggestion is not feasible save in areas with concentrations of the group.

# 11

# IMPROVING INTERETHNIC RELATIONSHIPS: HOW EFFECTIVE IS COOPERATION?

Jan Pieter van Oudenhoven
Groningen University
The Netherlands

The social psychological literature does not offer an optimistic picture of interethnic relations. Researchers have repeatedly established the occurrence and perseverance of prejudices and discrimination, and social psychological theories seem to be more successful in explaining how and why prejudices come about than in indicating how they could be changed.

Yet there seems to be one exception. Both theories and empirical evidence point to cooperation as a strategy for the reduction of intergroup conflict. In a cooperative goal structure the goals of the separate individuals or subgroups are linked together in such a way that an individual (or subgroup) can attain his or her goal only if the other participants can attain their goals as well. Admittedly, cooperation has primarily been used as an instrument for improving the interpersonal relationships between members of different ethnic groups within concrete settings such as the classroom, the neighborhood or the workplace. Not much attention has yet been paid to the generalization of improved personal relationships to the different ethnic groups in general. As a matter of fact, the question of how to improve relationships

between various ethnic groups in general is an important problem in most societies.

In the next section of this chapter we will describe what relevant
social psychological theories for intergroup relations (would) conclude with respect to cooperation. Subsequently, we will discuss
research on cooperation in the classroom and some methods of
cooperative learning, since it is in schools that most studies on cooperation have been done. In the final section of this chapter we will
deal with some questions that deserve further research. These questions refer to the generalization problem and the conditions under
which cooperation may be effective.


THEORIES ON COOPERATION

Of all theories of intergroup relations the *contact hypothesis* is most
directed toward prejudice reduction. It is a theory of change whereas most other theories primarily try to explain the origin of prejudice and related phenomena. In the contact hypothesis, as formulated by Allport (1954), cooperative interdependence was mentioned
as an important condition for the reduction of ethnic prejudice:
"Prejudice may be reduced by equal status contact between majority
and minority groups in the pursuit of common goals. The effect is
greatly enhanced if this contact is sanctioned by institutional supports (i.e. by law, custom, or local atmosphere), and provided it is of
a sort that leads to the perception of common interests and common
humanity between members of the two groups" (Allport, 1954,
p.281).
In more recent formulations of the contact hypothesis (e.g. Amir,
1969; Cook, 1978) cooperative interdependence remains of central
importance. According to Slavin (1985), who applies cooperation to
school settings, cooperative learning methods may even satisfy all
the conditions outlined by Allport for positive effects of desegregation on race relations. Most cooperative learning methods are structured to give each student a chance to make a substantial contribution to the team, so that teammates will be equal – at least in the
sense of role equality. Cooperative learning offers a daily opportunity for intense interpersonal contact among students of different
races. And finally, when the teacher assigns students of different
racial or ethnic backgrounds to work together, this implies un-

equivocal support on the teacher's part for the idea that interracial or interethnic interaction is officially sanctioned.

According to the *belief congruence theory* (Rokeach, 1960), similarity between individuals' belief systems is an important determinant of their attitudes towards each other. We are attracted to persons with similar beliefs since they tend to validate our own. Blacks are discriminated against not because they are black, but because they are perceived to hold different attitudes from those who discriminate against them. Several studies have shown that white subjects are usually more attracted to a black person with similar beliefs than to a white person with different beliefs (a.o. Byrne & Wong, 1968; Rokeach & Mezei, 1966). Apparently, their attitudes are more important than the color of their skin. The belief congruence approach assumes that the development of prejudice stems from the absence of sufficient information and/or the existence of erroneous information held by one group about the other. Contact between members of two groups facilitates positive attitude change towards the other group because it provides the opportunity to find out that the other group's beliefs are less peculiar than expected. Contact must occur under positive circumstances. Obviously, cooperation may offer a good opportunity for positive contact. The theory makes one important assumption: namely, that it is a misperception that the beliefs of ingroup and outgroup are dissimilar. However, this assumption does not always apply; differences in beliefs between ingroup and outgroup may be real and in some cases even increase after the contact has occurred.

In the early sixties Sherif developed his *functional theory of intergroup behavior* on the basis of his classic summer camp studies. In a well-known experiment the Sherifs and their colleagues (Sherif, Harvey, White, Hood & Sherif, 1961) showed that hostility decreases when rival groups work together to achieve a mutually desired goal. Boys (11 to 12 years old) at a summer camp were first divided into two groups. None of the boys knew each other before coming to the camp. In this first stage each group gradually developed their own role differentiations and social norms. In the next stage, the camp leaders (the researchers) organized competitive activities between the groups which led to overt hostility both within and outside the contests. In the final stage the rival groups were provided with superordinate goals: goals desired by both parties but which neither could achieve without the help of the other. As the boys worked together to master difficulties facing them, the conflict between them decreased and they even developed some liking for each other. New friendships developed, cutting across group boundaries.

The conclusion Sherif (1967) draws from these observations is that when conflicting groups meet under conditions of superordinate goals, cooperative activity towards the goal has a positive effect on intergroup relations. It reduces social distance, decreases hostile out-group attitudes and stereotypes, and makes future intergroup conflict less likely. Sherif's research has had an enormous influence upon theories of intergroup relations, particularly upon the contact hypothesis. However, not all social psychologists agree entirely with his conclusion about the effects of cooperative activity.

Worchel, Andreoli and Folger (1977) speculated that the successful outcomes of the cooperative efforts rather than the cooperation itself may have led to reduced hostilities. To test this hypothesis they designed an experiment in which groups working together either failed or succeeded. When cooperation between groups that had been competing on two previous tasks ended in failure, the groups decreased their liking of the outgroup. After success, however, all groups – even the groups that had been competing before – liked the outgroup members more than before they did the cooperative task. Apparently, success may be an important factor in the early phases of intergroup cooperation, particularly when the social climate is still conflicting or competitive. Interestingly, groups that had cooperated on two previous tasks did increase their liking of the outgroup, even if they failed on the last task.

The functional role of cooperation for the improvement of inter-ethnic relations is further disputed by protagonists of *social identity theory*, in particular by Turner. Social identity theory (Tajfel & Turner, 1979; 1985) essentially contends that people seek a positive social identity. Since part of our identity is defined in terms of group affiliations, it follows that a positive social identity requires that one's own group be positively distinctive from relevant comparison groups. The search for favorable distinction from other groups leads to ingroup favoritism and it may also lead to ingroup bias. According to Turner (1981, p.66), ingroup favoritism is any tendency to favor ingroup over outgroup members on perceptual, attitudinal or behavioral dimensions. Ingroup bias is an unreasonable or unjustifiable form of ingroup favoritism, in that it goes beyond the objective evidence or requirements of the situation as, for example, derogatory outgroup attitudes which have no veridical basis or discriminatory intergroup behavior which does not directly benefit ingroup members (Turner, 1981, p.66). A considerable amount of research shows that when individuals categorize others as members of a particular group they tend to perceive them in stereotyped ways and to react in a depersonalized manner (e.g. Brewer & Miller, 1984; Deschamps & Doise, 1978). The need for a positive social identity may enhance the salience of ethnic boundaries and consequently evoke social competition and rejection between groups.

Can intergroup conflict be overcome by cooperation according to social identity theory? Turner (1981, p. 74-75) states that competition between groups tends to develop into mutual hostility accompanied by ingroup-outgroup biases; and that cooperative intergroup interaction is conducive to a decrease in social distance and a reduction of ingroup-outgroup biases. However, Turner also holds that the different effects of cooperation and competition may disappear when social interaction is minimized and the salience of the ingroup-outgroup division maximized. Under these circumstances both competition and cooperation lead to ingroup-outgroup biases. It is doubtful, however, whether a situation with minimal interaction and maximized ingroup-outgroup divisions can be called a cooperative situation at all. Turner concludes that it is the formation of a superordinate group and not the superordination of objective interests that helps to overcome intergroup divisions. The feeling of belonging to a common group or category makes people willing to cooperate. Consequently, the best strategy for conflict-resolution would be to attempt to minimize or, if possible, to eliminate ingroup-outgroup distinctions directly. This can be accomplished by creating superordinate social identifications (e.g.: "We are all laborers!"), which should tend to produce cohesiveness between the conflicting groups (e.g.: strikers and non-strikers) and to the perception of cooperative interests between the members of the former subgroups as a direct consequence (Turner, p.99).

Theoretically, Turner may be right that it is the perception of belonging to a superordinate group that leads to cooperative behavior. However, while it may be feasible with experimental groups to form a superordinate group without common interests, it is extremely difficult to do so with existing groups that differ in religion, race or language. In practice, having common interests will virtually always be coupled with the perception of belonging to a superordinate group. There are nevertheless several other ways in which the salience of social categories can be reduced or the negative consequences of social categorization mitigated. For example, Deschamps and Doise (1978) have shown that the existence of 'cross-cutting' categorical distinctions can mitigate the tendency toward intergroup discrimination. In fact, the formation of groups of heterogeneous composition with respect to ethnicity and gender as part of most methods of cooperative learning is a way of cutting across social categories.

Turner's analysis is not the only elaboration on social identity theory. Brewer and Miller (1984), for instance, who like Turner base their research on social identity theory, contend that it is cooperative interdependence that promotes decreased social categorization instead of the other way round. It is more realistic to assume that cooperative interdependence and reduced social categorization influence each other in both directions: When two or more persons have positively interdependent goals, then possible social categorizations distinguishing them will become less salient (Brewer & Miller's point of view); on the other hand, when two or more persons, for instance from different ethnic backgrounds, are categorized as belonging to the same group (e.g. Pennsylvanians, dentists, Democrats), they will probably have a cooperative orientation and behave accordingly (Turner's viewpoint).

In his *theory of intergroup attributions* Hewstone (in this volume) points out that the role of attributions is crucial in the development, maintenance and reduction of intergroup conflict. Intergroup attribution refers to how members of different social groups explain the behavior and its outcomes of their own and other groups. The theory is not focused on the explanation of behavior of individuals as such, but on the explanation of the behavior of individuals as members of social groups. In particular, Hewstone highlights the role of cognitive processes which inhibit the generalization of experiences of positive interethnic contact to other outgroup members. An example is the tendency to attribute the success of an ingroup member to ability and that of an outgroup member to an external factor such as luck. Since – according to the theory – there generally exist negative expectancies for outgroup behavior, unexpected positive behavior of outgroup members will be attributed to external factors. This tendency not to acknowledge personal motives for positive outgroup behavior makes it difficult to bring about a favorable change in troublesome intergroup relations. But even when it is attributed internally, the unexpected positive behavior can be 'explained away' by treating the individual as 'a special case', which means that it has little impact on the outgroup stereotype. If, and only if, the individual is seen as a typical outgroup member, is there a real chance of generalized change of outgroup attitudes.

Support for this idea is found in an experiment by Wilder (1984) in which the representativeness of an outgroup member and the pleasantness of her behavior were manipulated. Subjects were asked to perform a cooperative task with a student from a rival college. The outgroup member behaved either in a pleasant and supportive way or in an unpleasant and critical fashion, and she presented herself as either a typical or an atypical outgroup member. The results of the experiment showed a significant improvement in the evaluation of the outgroup as a whole only when the encounter was pleasant and the outgroup member could be seen as typical.

Hewstone does not refer to cooperation in his chapter on intergroup attribution, but elsewhere (Hewstone & Brown, 1986) it is mentioned as an apparently successful technique to reduce interethnic tensions. But Hewstone and Brown doubt whether cooperation will have much effect on groups living in contexts that strengthen intergroup distinctions and whose differences are insurmountable. When people from different groups cooperate, however, they have the opportunity to notice the effort exerted by their co-workers and are less likely to attribute externally an unexpected positive contribution of an outgroup member. Cooperation may thus counteract to some extent the conflict-sustaining influence of external attributions for positive outgroup behavior.

Generally, the theoretical approaches discussed above indicate that cooperation may be helpful to improve interethnic relationships under specific conditions. Cooperation should preferably be successful. A theoretical as well as practical problem is whether attitude change about outgroup members will generalize to outgroup members outside the cooperating group. We will return to this problem in the next section when we discuss some applied research in the field of intergroup cooperation, specifically in educational settings, since that is where enough studies have been realized to enable empirically based conclusions to be drawn about the conditions under which cooperation leads to more favorable interethnic relations.

COOPERATION IN THE CLASSROOM: METHODOLOGY AND RESEARCH

Social psychologists have greatly influenced the policy of school desegregation. Their research evidence was used by the Supreme Court of the United States to decide in favor of desegregation in 1954 (Oskamp, 1984). Since then, many researchers have conducted studies on the processes and effects of desegregation. Longitudinal studies have generally shown small or nonexistent achievement gains for minority students, occasional decreases in their self-esteem, and fairly frequent increases in racial tensions and prejudice between whites and minorities as consequences of desegregation programs (Oskamp, 1984). The assumption that relationships between blacks and whites would improve by merely bringing them together in one school was rather naive indeed. However, during the last decade researchers have turned their efforts toward developing a more effective process of desegregation. The development of cooperative learning has undoubtedly been the most successful effort. An increasing number of studies suggest that cooperative learning generally leads to improved intergroup relations (for an overview see Johnson, Johnson & Maruyama, 1984; Sharan, 1980; Slavin, 1985). In this section we will briefly describe the sort of cooperative methods used in desegregated schools and discuss two important issues: the status of the participating groups and the generalization of a positive attitude change to the outgroup members that do not participate in the cooperation.

*Methods of Cooperative Learning*

Of the rapidly growing number of cooperative methods, we will discuss only the most widely used, and our own method of Dyadic Peer Learning because of its relatively easy applicability. Common to all methods is the division of the class into small teams. Students in each team are made positively interdependent by a form of group reward or the structure of the task or a combination of both principles. The most well-known methods are: Student Teams-Achievement Divisions, Teams-Games-Tournament, Jigsaw, Jigsaw II, Group Investigation, and Learning Together (for a review see Kagan, 1985; Slavin, 1985).

*Student Teams-Achievement Divisions* (Slavin, 1983) is a peer tutoring technique. The teams are composed of four students who are carefully selected to represent a cross section of the class; the teams are heterogeneous with regard to gender, ability, and ethnic background. The group members study materials taught previously by the teacher or in an audiovisual presentation. Most often, the team members quiz each other, working from worksheets that consist of problems and information to be mastered. All students in the class take individual tests, and team scores are computed on the basis of the degree to which each student improves on his or her own past record. These group scores are published in weekly newsletters. The group members are primarily rewarded on the basis of their group's score, but recognition is also provided for individuals who perform exceptionally well.

*Teams-Games-Tournaments* is essentially the same as Student Teams-Achievement Divisions in method but it replaces the quizzes and improvement score system used in Student Teams-Achievement Divisions with a system of tournaments. The students play games in which they win points by demonstrating knowledge of the academic material which has been practiced in teams. Students are insured of having an equal opportunity to earn points for their teams. Students representing a team compete with students of comparable past performance from other teams to try to contribute to their team scores.

In *Jigsaw* (Aronson, Blaney, Stephan, Sikes & Snapp, 1978) students are assigned to teams consisting of about six members. Teams are heterogeneous with regard to gender, ability, ethnic background and personality factors such as assertiveness. Each team member is provided with a unique set of learning material on an academic unit, such as Latin America. One student may be appointed as a specialist on Latin America's history, a second student on its economy, another on its culture and so on. In a sense, each student on a learning team has but one piece of a jigsaw puzzle. The students read their material and then discuss it in 'expert groups' consisting of students who have the same information but come from different teams. The specialists then return to their own groups to teach their group mates. All students are tested on how well they master the whole unit, and receive individual grades; there is no group reward. Because communication among team members is an essential part of Jigsaw, special team-building activities are included to prepare the students to cooperate. In this method there is not only cooperation within the teams, but also between members of different teams, as the expert groups have to communicate cooperatively as well.

*Jigsaw II* is a modification of Jigsaw by Slavin (1980), according to whom group rewarding is the main principle of cooperative learning. In Jigsaw II students discuss their topics in expert groups and teach them to their group mates as in original Jigsaw. The main differences are: teams have four or five members instead of six; all students read the whole learning unit, but each member is given an individual topic on which to become an expert; individual quiz scores in Jigsaw II are added up to form team scores, and teams are recognized in a class newsletter as in Student Teams-Achievement Divisions. Because all students have access to all learning materials, interdependence between students is lessened. However, the use of existing learning materials makes Jigsaw II more practical.

*Group Investigation* (Sharan & Hertz-Lazarowitz, 1980) was designed to provide students with very broad and diverse experiences, in contrast with the above-mentioned techniques, which are oriented toward acquisition of facts and skills. Students form their own groups consisting of three to six members. Ideally, the groups are composed of students of both sexes, with different abilities, and from varying ethnic backgrounds. Obviously, the students' free choice of groups and the ethnic, gender and ability-level heterogeneity within the groups are often incompatible goals. The teacher tries to reconcile these goals with discussion, which seems to be no easy task. The method involves the students in planning study activities, carrying out these plans through mutual assistance and exchange, participating in small-group discussions, and preparing group products. Each group has to choose a subtopic from a unit being studied by the entire class, further divide their group task into individual tasks, and prepare a group report. The group makes a presentation (e.g. an exhibition or a debate) to communicate its findings to the entire class and is evaluated on the basis of the quality of this report.

*Learning Together* is the name for a series of cooperative learning methods developed by Johnson and Johnson (1975), in which students work in small groups to complete a common task. The groups are heterogeneous with respect to gender, ability and ethnic background. Essential to the method is that the students experience a sort of controversy. Controversy exists when one student's ideas are incompatible with those of another. Controversy creates academic curiosity and results in students reconceptualizing what they know about the subject studied. Although the effects of controversy refer primarily to cognitive variables, there seems to be evidence that it promotes liking among participants as well (Johnson, Johnson & Smith, 1986). The students are praised and rewarded as a group. The Johnson methods are closest to pure cooperation as they do not contain individual and/or competitive elements, which are clearly present in methods like Student Teams-Achievement Divisions and Teams-Games-Tournaments.

In our method of *Dyadic Peer Learning* (Van Oudenhoven, Van Berkum & Swen-Koopmans, 1987) students have to do assignments in peer dyads, but receive individual feedback. The individuals in each dyad have to wait for each other to finish each exercise, correct each other's work, and discuss each other's mistakes. Assignment to the peer dyads is based on the scores of an achievement test. On the basis of these scores the class is divided into four levels. The highest scoring students from the first level are paired with the highest from the second level. The same is done with students from levels three and four. Thus, achievement differences in each dyad are never disparate, so that the problem of status inequality is avoided. Assignment of students of different gender or ethnicity to the dyads occurs on a rather random basis as a consequence of this procedure. No research on interethnic relations has yet been done using this method.

The methods differ in the degree of cooperation involved. Learning Together and Jigsaw do not stimulate competition at all. Teams-Games-Tournament and Student Teams-Achievement Divisions, on the other hand, clearly have some competitive elements as teams have to compete. Intergroup competition may be a mixed blessing.

Although it is a well-established fact that intergroup competition leads to more within-group cohesion (e.g. De Vries, Edwards & Slavin, 1978), it is equally well-known that it leads to negative feelings towards the outgroup. De Vries *et al.* therefore advise that groups of cooperating students frequently change their membership composition. In fact, their suggestion is to apply periodically what Deschamps and Doise (1978) would call cross-cutting categorical distinctions.

Apart from that, it may still be possible to have students working together in small groups and obtain positive cross-ethnic relationships with students that do not belong to the own group. This is the conclusion from a study by Johnson and Johnson (1985), who investigated the effects of intergroup competition and cooperation on the relationships between black and white students. Within both conditions, students were assigned to groups of four on a stratified random basis so that the groups were comparable with regard to ability level, race, and gender. In the intergroup-cooperation condition, the emphasis was placed on how well the entire class achieved. Students were instructed first to learn the assigned material, then to ensure that all members of their group knew the material, and finally to assist other groups until everyone in the class learned the assignments. Daily feedback was given on how well the entire class was doing. For each day that the whole class met the criteria for excellence, the class received an award. In the intergroup-competition condition, the emphasis was placed on which group achieved the best performance. Students were instructed first to learn the assigned material and then to ensure that their group mates knew the material. Daily feedback was provided on which group was ahead in the competition. All the members of the winning group received a prize at the end of the ten-day study; the losing groups received nothing. Intergroup cooperation promoted more cross-ethnic interpersonal attraction than did intergroup competition. In addition, the students under the cooperative condition perceived less friction and more collaboration among group members. Both intergroup cooperation and intergroup competition led to more favorable interethnic attitudes than individual instruction.

Interestingly, the minority students in particular reacted favorably to the cooperative group experiences. They seem to have responded positively to the support inherent in cooperative learning situations, as Johnson and Johnson (1985) argue, because minority students may be relatively deprived of constructive peer relationships within most learning situations. This speculation is in agreement with Kagan's thesis (Kagan, 1980) regarding structural bias in classroom instruction, which states that some teaching methods exert negative effects on pupils from certain ethnic groups. This thesis emerged from Kagan's reseach on the cooperative-competitive behavior of Mexican-American children as a function of their cultural history. Mexican-American children display a more cooperative orientation than their Anglo-American peers and have not been socialized to be competitive. Therefore, it is reasonable to assume that traditional whole-class instruction, with its competitive orientation and its use of social comparison as a major strategy for arousing pupils' motivation to learn, would exert a negative effect on the behavior of Mexican-American children in school settings. Kagan's thesis is supported by anthropological research; several anthropologists (e.g. Whiting & Whiting, 1975) found that non-western cultures are generally more cooperatively oriented than western cultures.

*The Status Problem*

There is a general consensus that in order to achieve positive interethnic attitudes within cooperating teams the collaborating group members should be equal in status (Allport, 1954; Bizman & Amir, 1984; Cook, 1969). It is also widely accepted that equal status within the contact situation (e.g. a student's popularity or academic achievement) is more important for attitudinal change than equality of status on general factors, such as economic status or the status based on ethnic origin (Bizman & Amir, 1984). This equal-status condition may be fulfilled in cooperative learning through equality of roles (Slavin, 1985). The impressive number of studies reporting positive interethnic attitude shifts as a consequence of cooperation suggest that differences in status based on ethnic background are by no means insurmountable.

Yet Cohen (1980) is rather pessimistic about the feasibility of creating equal status between majority and minority group members. She argues that, in the US, race operates as a diffuse status characteristic that triggers certain expectations regarding the competence of the persons holding that status. Specifically, her research demonstrates that even when the black and white students participating in an interaction are carefully matched on status-relevant variables, including social class, the white students tend to be more active and more influential (Cohen, 1972). Cohen does not deny the successful results of many cooperative intervention strategies, but she argues that the success of a program does not depend on positive affective changes alone. Liking does not necessarily indicate respect or compensate for lack of it. Attitudes about black intellectual incompetence are not changed by an increased liking.

Cohen makes an important point distinguishing between liking and respect for competence, but she probably overestimates the influence of general status characteristics in the classroom. According to the expectation states theory, specific situational variables seem to become more important in stabilized task situations (Berger, Wagner & Zelditch, 1985). Cohen's analysis may be more applicable to educational settings, such as whole-class instruction, in which social comparison on academic competence is very salient. Cooperative learning, on the contrary, aims to reduce processes of social comparison based on academic achievement or competition for grades. And several cooperative learning methods that still use intergroup competition on academic achievement offer the opportunity for low achievers to experience success by some kind of individualized feedback. For instance, the scores students contribute to their teams in Student Teams-Achievement Divisions are based on the students' improvement on their own level of past performance. In Dyadic Peer Learning achievement differences within the dyads are relatively small.

Cooperative learning methods reduce the salience of social comparison processes, but do not prevent students from perceiving the differences between high and low achievers. The main difference from individual goal structure instruction is that in cooperative learning individual differences in academic achievement are stressed less. This means that low achievers are not discouraged as much as they usually are in whole-class instruction by the continuous explicit feedback that they are performing less well than their fellow students.

Not only does cooperative learning reduce the salience of social comparison processes, it also helps – at least to a certain degree – to overcome the achievement gap between minority and majority students since cooperative learning tends to be particularly effective for minority students' performance (a.o. Kagan, 1980; Oskamp, 1984). Consequently, the problem of minority students' lower competence may be partly resolved by the process of cooperative learning.

## Generalization of Attitude Change

Although cooperation between members of different ethnic groups usually leads to more favorable attitudes towards the persons involved in the cooperation, there is serious theoretical and empirical doubt about the generalization of those attitude changes to the non-participating outgroup members. As mentioned above, Hewstone (this book), in his attributional analysis of intergroup relations, states that generalization of attitude change will only take place when the outgroup member is seen as a *typical* member of the outgroup. This implies that group membership must be made visible and that outgroup members should not be seen as exceptions to the rule. Support for this idea was found in the experiment by Wilder (1984), to whom we referred earlier.

Brewer and Miller (1984), however, present a point of view to explain the effects of cooperation on intergroup relations that is different from that of Hewstone, in spite of the fact that they too have based their approach on Tajfel's social identity theory. They state that interaction should be interpersonally oriented and not category-based to improve intergroup relations. According to their theoretical model, *decreased* social categorization, instead of being the result of positive intergroup interaction, is an important intermediating variable to produce interpersonal attraction. Cooperative goals provide an opportunity for reducing the salience of category membership, but whether they do, will depend on the task structure and the nature of the interaction the goals promote among team members. The environment should be socially oriented rather than task-oriented and the structure of the contact should promote interpersonal orientation and open communication. Brewer and Miller therefore consider interteam competition a less suitable cooperative strategy since it may have the effect of enhancing task focus and thus lead team members to ignore personalizing information because of its irrelevance to task objectives. As a consequence, such cooperation provides less opportunity to acquire information that would counteract social categorization.

Brewer and Miller rightly say that the effects of intrateam cooperation on acceptance of outgroup members not involved in the cooperative venture will be reduced under conditions of intergroup competition. The positive effect on attitudes within the team is indeed difficult to generalize to persons outside the group when there is still competition with outgroup members. In our opinion, it is the competitive character of the interaction and not the perception of social categories that makes interteam competition a less suitable instrument for attitude change toward the outgroup as a whole. However, we agree with Brewer and Miller that social categorisation should not be too salient when minority members are less competent in performing the task and it is certainly true that competition tends to make task-relevant differences more salient. Most cooperative learning methods use some stratified assignment of ethnic students to the teams in order to obtain heterogeneous teams. Such a procedure, however, may make ethnicity a salient feature. Dyadic Peer Learning avoids any salience of ethnic categories because dyads are not formed on the basis of ethnicity, but on the basis of achievement. There are several methods of cooperative learning that fulfill Brewer and Miller's requirement of non-competition between groups. Jigsaw, Group Investigation, Learning Together and Dyadic Peer Learning are all methods that do not use competition between groups. The method of intergroup cooperation (Johnson and Johnson, 1985), which we described earlier, explicitly stimulates interteam cooperation.

Brewer and Miller are by no means the only researchers who have stressed the importance of personal contact. Other theorists like Allport (1954), Amir (1969) and Cook (1969) have argued as well that fairly close interpersonal contact is more likely to create positive feelings than more distant contact, since the high 'acquaintance potential' of the former situation increases the opportunity for individuals to discover that they have similar interests, attitudes and the like.

The problem of how to bring about generalization of a positive attitude shift to the outgroup as a whole seems to be paradoxical. On the one hand, the fading of category boundaries through interpersonal interactions helps to improve relationships between the cooperating members of different groups. On the other hand, blurred category boundaries make it less likely that positive effects generalize to other representatives of the outgroup that are not involved in the cooperation. It is doubtful whether this paradox can be solved, but perhaps the best way to handle it is to minimize social categorization when initiating and supporting the process of cooperation. In a later stage, when the cooperation process has been well established, the social categories to which the participants belong may gradually be made more visible.

In this section we have presented cooperative learning as a method to improve intergroup relations and described the most widely used methods of cooperative learning. Subsequently, we discussed the problems of the status of team members and the generalization of attitude change to other outgroup members. In the concluding section we will briefly provide a further comment on these problems in our general discussion of cooperation as a strategy for the reduction of intergroup conflict.

CONCLUSIONS

Most social psychological theories of intergroup relations suggest cooperation as an important factor that may reduce intergroup conflict. In spite of this almost general consensus, social psychologists still vary considerably in their explanations of the effects of cooperation, and in the description of the conditions under which cooperation may be effective.

With respect to the explanations of the effects of cooperation, the points of view of Turner (1981) and Brewer and Miller (1984) in particular deserve further research. According to Turner, it is the perception of belonging to a superordinate group that leads to cooperative behavior and not the formation of common interests that produces a reduction of social categorization. Brewer and Miller, on the other hand, state that cooperative interdependence reduces the perception of social categories. A decrease in social categorization in turn leads to increased positive interpersonal interaction. As we suggested earlier, causality is probably bidirectional, which means that both points of view may be correct to some extent. Brewer and Miller compare their model with two alternative models, one in which decreased social categorization is seen as the end product of increased positive intergroup interaction provoked by cooperative interdependence. In the other model cooperative interdependence is viewed as a direct cause of a whole set of dependent variables among which figure decreased social categorization and increased positive intergroup interaction. It is not yet possible to choose one model or another on the basis of the available research since most studies on cooperation in the classroom until now have been designed to measure the effects of various cooperative learning methods and only marginally to study the intermediating processes. Moreover, the implementation of cooperation has quite often been confounded with some form of (decrease in) social categorization, which makes it very hard to give an answer to the theoretical questions raised by Turner and Brewer and Miller. Their alternative models offer a challenging research program for the current generation of both experimental and field researchers. The question of whether cooperation eliminates social categorization, or whether reduction of social categorization leads to the perception of common interests is theoretically interesting but it is a clear and important implementation question as well. For practical reasons it is important to know whether to focus on the creation of common interests or on the building of a common group. As long as this question remains unanswered, it will not do any harm to combine both a reduction of social categorization and an induction of common interests in implementation programs.

With respect to the conditions under which cooperation may be effective, the following three questions seem to be particularly important and need to be further investigated:

1 How to realize equality of status? To this question our provisional answer would be: by realizing equality of role behavior in combination with a reduction of the salience of the status characteristic on which ingroup and outgroup vary.

2 How to bring about a generalization of positive attitude change to other members of the outgroup? With regard to this second question, it is more difficult to give a clear answer since studies on cooperative learning have primarily focused on attitudes and friendships with fellow students in the same classroom. On the basis of the available evidence from experimental research, however, we tend to agree with Hewstone that members of the outgroup should be perceived as more or less representative for their group in order for generalization of attitude change to take place. There is, however, a political reason as well why one should not

ignore intergroup differences. Most ethnic groups foster more or less their social and cultural identity and would strongly resist giving up or repressing this identity in order to make the conflicting groups less distinctive. On the other hand, those who favor a pluralist approach to desegregation have to take some common identification into account (for a discussion of pluralism see chapter 15). Schofield (1986) warns that schools adopting a pluralistic perspective, which tends to make group membership relatively salient, need to find ways of minimizing the potentially negative impact of this stance on intergroup relations.

3 Is success a necessary condition for cooperation to be effective? Not necessarily for groups that have a tradition of collaborating. However, cooperation should preferably be successful in new groups, particularly if the constituent subgroups have a history of conflicts. As we have seen above, some cooperative learning methods like Student Teams-Achievement Divisions, Teams-Games-Tournaments, and Jigsaw II are designed in a way that all group members are able to experience success.

Research on cooperative learning has demonstrated that cooperation is a useful strategy for the improvement of intergroup relations. Other strategies to improve interethnic relationships seem to be much less successful (for an evaluation of several strategies see chapter 15). Much of the research on cooperation, however, lacks theoretical precision since it arose out of an urgent need to compensate for the disappointing results of school desegregation. Considering the importance of cooperation for theoreticians as a concept and for practitioners as a tool, we may be optimistic about a further integration of theoretical and applied research on this topic, since important theoretical questions often have direct relevance for the implementation of cooperation. A good example is the problem of generalization of attitude change to the outgroup as a whole.

Most of the studies on cooperation have been done in school settings. That is not surprising since the school constitutes an important socializing agent which offers an opportunity to bring about positive intergroup relations before prejudice has become too firmly established. Moreover, as Amir, Sharan and Ben-Ari (1984) argue, a school is the only institution that encompasses the entire population of a nation's children. According to them, the school is also the preferred setting for integration in a country of immigrants, such as

Israel, because at school children with parents from different countries and cultures are more similar to each other and less socially distant from their peers than are their respective parents in their daily lives. This statement also applies to other immigration countries such as Canada and the United States. While schools may be very important, the time has come to do more cooperation research in other settings than schools as well, such as the workplace, the neighborhood and leisure activities.

There is by now sufficient research on desegregation to conclude that contact alone is not enough to improve relations between ethnic groups. Intergroup contact can help to improve intergroup relations but only when several conditions are met. Since one of the main conditions is the creation of common goals, it is better to re-baptize the 'contact hypothesis' into the 'cooperation hypothesis'. One advantage would be that no more energy would have to be wasted on the permanently recurring refutation of the simple interpretation of the contact hypothesis: i.e. that contact alone would be sufficient to improve intergroup relations. How much more fruitful it would be to spend that energy on research concerning the conditions under which cooperation leads to the improvement of interethnic relationships.

ACKNOWLEDGEMENTS

The author would like to thank Naomi Ellemers, Ad Van Knippenberg and Bert Wiersema for their critical comments on an earlier version of this chapter.

# 12

# THE CULTURE ASSIMILATOR: IS IT POSSIBLE TO IMPROVE INTERETHNIC RELATIONS BY EMPHASIZING ETHNIC DIFFERENCES?

Henriëtte van den Heuvel
Roel W. Meertens*
University of Amsterdam
The Netherlands

Since the Second World War, the social psychological approach to intergroup conflict and interethnic relations has been dominated by the so-called Contact Hypothesis (Allport, 1954; Cook, 1978; Pettigrew, 1971). According to this hypothesis, face-to-face contact between members of different (ethnic) groups will lead to a positive change in the attitude and behavior of members of the ingroup with regard to members of the outgroup. Research inspired by the Contact Hypothesis shows mixed results. Positive effects of contact have been reported, as well as negative effects, depending on the conditions of contact (Amir, 1976; Hewstone, 1988; Hewstone & Brown, 1986; Stephan & Stephan, 1984). A promising procedure, meant to improve intergroup contact, will be described in this chapter: the Culture Assimilator training procedure. This procedure does not attempt, however, to influence interethnic contact directly, but to influence the expectations about contacts, and thereby indirectly those contacts themselves.

We start with an outline of the principles of the Culture Assimilator, and present a short overview of results obtained with the procedure. In the next section, we look at the Culture Assimilator from the point of view of more recent theorizing and research on intergroup conflict and interethnic relations, as summarized by Hewstone (this volume). Our focus will be on emphasizing differences and/or similarities between groups. In the last section, we present guidelines for the application of the Culture Assimilator procedure in interethnic relations.


## GENERAL PRINCIPLES OF THE CULTURE ASSIMILATOR

Essentially the Culture Assimilator " ... is a programmed learning experience designed to expose members of one culture to some of the basic concepts, attitudes, role perceptions, customs, and values of another culture" (Fiedler, Mitchell & Triandis, 1971, p.95). The first Culture Assimilators were developed for Americans going temporarily abroad to work or study in a foreign culture. The subjects in the training procedure are presented with about one hundred items. Each item starts with a description of an intercultural, interethnic, or intergroup interaction incident between two persons, one a member of group A, the other a member of group B. The description of the interaction contains some degree of interpersonal ignorance, misunderstanding, or difficulty. Next, four different interpretations of the incident are presented to the subject, a member of group A. These interpretations are possible explanations for, or attributions of, the behavior of the member of group B during the described interaction. Three of these attributions are incorrect from the point of view of group B, although they are plausible for members of group A. Only one attribution is correct from the perspective of group B. The subjects have to select one of these four alternative interpretations. If they have chosen the correct alternative, they receive positive feedback and are informed why that answer is indeed the correct one, according to group B. If they have chosen an incorrect alternative, they are told so, receive mild criticism, and are instructed to try again.

During the training procedure, the subjects continuously receive feedback on the correctness of their attributions of the behavior of members of group B. Gradually, they learn to explain that behavior less in terms of their own culture, and more and more as the members of the other culture would do; in the words of Triandis (1975), they learn to make 'isomorphic attributions'. Triandis (1975) lists the different types of information about the other culture which are in principle provided by the training procedure. These pieces of information form the often unrecognized, implicit value premises, roles, norms, and attitudes of a group, or, as Triandis labels it, the subjective culture of that group. Briefly, this information concerns: (a) norms for different kinds of situations, (b) different role perceptions between the two cultures, (c) the links between general intentions and specific behaviors, (d) frequently found self-concepts, (e) valued and disvalued behaviors, with their frequently associated antecedents and consequents, (f) the (relative) influence of norms, roles, self-concept, general intentions and affect toward the behavior, and (g) the kinds of reinforcements that are expected in different situations (Triandis, 1975, p.69).

As a result of the training procedure, the subjects acquire a greater understanding of the other culture: They learn to make isomorphic attributions, and therefore develop more appropriate expectancies about the behavior of the members of the other group. Consequently, subjects should be better able to predict the behavior of their interaction partners and should experience less anxiety during actual interaction with members of the other group. They should also develop a more positive attitude, which in turn should lead to more positive and mutually reinforcing behavior, with less ignorance, difficulties and misunderstanding. Box 1 gives an example of a Culture Assimilator item.

A Culture Assimilator item from Brislin *et al.* (1986, page 154).

*Tom Bancroft, the top salesman of his midwestern U.S. area, was asked to head up a presentation of his office equipment firm to a Latin American company. He had set up an appointment for the day he arrived, and even began explaining some of his objectives to the marketing representative sent to meet his plane. It seemed that the representative was always changing the subject and they persisted in asking lots of personal questions about Tom, his family, and interests. Tom was later informed that the meeting had been arranged for several days later, and his hosts hoped that he would be able to relax first and recover from his journey, perhaps see some sights and enjoy* hospitality. Tom responded by saying that he was quite fit and prepared to give a presentation that day, if possible. The representative seemed a little taken aback at this, but said he would discuss it with his superiors. Eventually, they agreed to meet with him, but at the subsequent meeting after chatting and some preliminaries, they suggested that since he might be tired they could continue the next day after he had some time to recover. During the next few days, Tom noticed that though they had said they wanted to discuss details of his presentation, they seemed to spend an inordinate amount of time on inconsequentials. This began to annoy Tom as he thought that the deal could have been closed several days ago. He just did not know what they were driving at. How would you help Tom view the situation?

1  The company was trying to check on Tom and his firm by finding out more information.
2  Latin Americans are not used to working hard and just wanted to relax more.
3  The Latin American Company was not really interested in the products of Tom's company and was just putting him off.
4  Tom's American perspective was concerned with getting the job done, whereas the Latin American company had the perspective of building a relationship with Tom and his company.

Feedback:
1  There are more efficient and accurate ways to find out such kinds of information about a company. This explanation is rather unlikely. There is a better answer. Please choose again.
2  This is a typical over-generalized stereotype of Latins. Although they do enjoy leisure time activities in their culture, they are also hard workers. Please choose again.
3  There is no evidence to support this. It is doubtful that they would waste so much time and money on Tom just to be kind while refusing him. There is a more reasonable answer.
4  This is the best answer. Much of American culture is concerned with efficiency and time. Although other cultures are also concerned about these things, Americans have a tendency to stress getting the job done, with emphasis on the end product rather than the means or the process in how it gets

*done. In Latin cultures although the product is impor-
tant, they are just as concerned about personal relationships and how a thing gets done.*

CONSTRUCTION OF A CULTURE ASSIMILATOR

Although differences exist in the specific domains and aspects that
are included in the various Assimilators that have been developed,
they all focus on 'critical problems'. That is, they emphasize those
differences between the key elements of the subjective cultures
involved that give rise to interpretational mistakes and interpersonal difficulties.

The first step in the development of a Culture Assimilator is sampling these critical problems. Albert (1983) lists several possibilities
for this sampling, such as observations, questionnaires, group discussions, and analysis of ethnographic and historical material. The
method most often used is to interview members of both cultures
involved who have participated in interethnic contact. The subjects
are asked to describe specific intercultural occurrences or events that
have had an effect on their attitude or behavior toward members of
the other culture (Fiedler, Mitchell and Triandis, 1971). The sampled
incidents are rewritten in the form of short episodes, and step two
consists of presenting these episodes to members of both cultures.
They are asked to explain the behavior of a member of the 'target
culture' that is described in the story. These explanations are stan-
dardized and are again presented to members of the ethnic groups
involved to establish which answers are 'correct' and which are
'incorrect'. On the basis of the information gathered in this process,
a Culture Assimilator is constructed.

SOME RESULTS OF CULTURE ASSIMILATOR TRAINING

Since the original development of the Culture Assimilator training
procedure, as described by Fiedler, Mitchell and Triandis (1971) and
Mitchell, Dossett, Fiedler and Triandis (1972), numerous assimilators
have been developed (see Albert, 1983; Brislin, Cushner, Cherrie &
Yong, 1986; Landis & Brislin, 1983; Stephan & Stephan, 1984). As
mentioned above, the first series of assimilators was developed for

Americans going temporarily abroad. For example, assimilators have been developed for Americans with respect to the cultures of Saudi Arabia, Thailand, Honduras, and Greece. Most of these programs were used for the training of Peace Corps volunteers and of military troops with overseas assignments. The general conclusions which can be drawn from these studies about the effects of the first series of assimilators are: (1) the Culture Assimilator procedure leads to an increase of isomorphic attributions, especially when the initial differences between the cultures are large; (2) this increase is accompanied by a somewhat more positive attitude toward the other group, and (3) by an improvement in the evaluation of the interpersonal relations; (4) however, the hoped-for changes in actual behavior with respect to members of the target group are rather weak (if they exist at all).

In this section, we will present in more detail some effect studies which form part of a second series of assimilators, especially developed for multicultural domestic use, like the relations between white and black Americans in the United States. We will limit our discussion to this second series because the purpose of this chapter is to explore the usefulness of the assimilator procedure with regard to different ethnic groups within one country, such as the ethnic minority groups of Turks, Moroccans and Surinamese in the Netherlands.

Weldon, Carlston, Rissman, Slobodin and Triandis (1975) were among the first who evaluated the Culture Assimilator as an instrument to improve the interaction between white and black Americans. The subjects of their study, 128 white male paid volunteers (university students), were divided into two groups: one group went through a Culture Assimilator training procedure that required 6-8 hours and mainly pertained to behavior with respect to hard-core unemployed blacks; the other group received no training. The dependent variables included attitude measurements and a Test of Intercultural Sensitivity (TICS) with respect to blacks. The TICS consists of a number of critical episodes with multiple choice answers. This test is in essence a parallel test of the Culture Assimilator. As in the Culture Assimilator, one answer is 'correct' according to the group in question, i.e. blacks. In addition to the testing procedure, the subjects worked together with black confederates of the experimenters on a complex performance task, which resulted in ratings by the confederates of the subjects on several behavioral dimensions. The subjects in the experimental condition showed differences from the subjects in the control condition on several dependent variables. The experimental subjects made more isomorphic attributions, described the behavior of the members of the target culture as more personally determined, and showed fewer signs of stereotypic thinking. Indirect measurements of attitude showed a somewhat more positive attitude toward blacks (although these results were somewhat equivocal). The ratings of the black confederates did not unequivocally show a preference for trained subjects over untrained subjects. In summary: the results showed (a) the expected positive effect of the Culture Assimilator training on cognitions; (b) but the effects on attitudes and on behavior (as rated by interacting blacks) were at best equivocal. Two possible explanations for these rather disappointing results are mentioned by Weldon *et al.* (1975): behavioral change following cognitive change may take more time than the subjects were permitted in this laboratory study, or the training may have failed to teach appropriate behaviors.

Landis, Day, McGrew, Thomas and Miller (1976) developed a Culture Assimilator procedure that pertained to interaction between white junior grade officers and black enlisted men in the American army. The effects were tested with 84 white junior grade officers from four army bases in the southern United States. The main dependent variable consisted of the TICS; no behavioral measurements were taken. The white officers showed a significant improvement on this variable, that is, they made more isomorphic attributions.

Landis, Brislin and Hulgus (1985) conducted a laboratory study to compare the effects of Culture Assimilator training by itself or in combination with behavioral contact. Subjects were 45 white male college students, assigned to five training conditions. We restrict our discussion to the results of the Culture Assimilator training. One dependent variable was again the TICS. A second dependent variable was the behavior of the subject in a role reversal situation, in which the subject was instructed to act as a black would do in front of a white confederate in three different situations. As a result of the Culture Assimilator procedure, the subjects made – again – more isomorphic attributions as measured by the TICS. However, the subjects could not assume the role of blacks better after the training than subjects in a control group, as scored by black judges. The Culture Assimilator training thus resulted in a cognitive change, but a behavioral change could not be established in this study. However, one could argue that the Culture Assimilator aims at changes in behavior of the whites as whites with respect to blacks. Behavior in a role reversal situation may not constitute a reasonable indicator of the trained behavior. The study also measured anxiety. Landis *et al.* (1985) supposed that the Culture Assimilator might lead to a greater anxiety in interethnic contact due to the challenge to long believed stereotypes. Therefore, it does not only take time, as Weldon *et al.* (1975) suggested, but also practice, to achieve a positive effect. This assumption appeared to be true: all subjects showed an increase in anxiety between pre- and post-test; however, the smallest increase was shown by subjects who had received behavioral training after having completed the Culture Assimilator training. Moreover, this group – in contrast to the group which had only completed the assimilator procedure – was seen as more 'personally likeable' by the black judges.

The conclusions, concerning these multicultural studies within one country parallel the conclusions of the evaluation of the first series of Culture Assimilator training results: the Culture Assimilator procedure leads to more isomorphic attributions, but attitudinal or behavioral changes toward members of the outgroup are less clear, or they do not occur at all without additional processing time and behavioral training (cf. Albert, 1983).

THEORETICAL CONSIDERATIONS

The question arises: Why are these changes in attributions not followed by changes in attitude and behavior? The literature on the Culture Assimilator (e.g. Albert, 1983) suggests mainly intra-psychic reasons: behavioral 'rehearsal' and attitude change takes more time than was permitted in the experimental studies; an embedded, negative attitude cannot be expected to change easily; or, the reverse, the attitudes of the subjects were already positive at the start of the training. Moreover, cognitive change and attitude change do not necessarily lead directly to behavioral change, according to the theories on attitude-behavior relations by Ajzen and Fishbein (1977) or Triandis (1980). For example, apart from the general attitude toward the object, other determinants of behavior are the attitude toward that specific behavior, and social norms or habits.

Beyond these intra-psychic reasons, other explanations result from structural societal relations between the groups involved, such as self-interest and discrimination. The somewhat more positive changes in attitude and behavior in the first series of Assimilators, as compared to the second series, may be explained in part by self-interest. The Americans going abroad had an apparent self-interest in good interethnic contacts, for they were dependent upon the other group for the success of their project. White Americans in the U.S.A., on the other hand, may not necessarily see the advantages of getting along well with blacks, for they are less dependent on them. They can continue to keep their distance, and there is no need for them to delve into the explanation for observed differences. Therefore, it may be sensible to restrict the development of Culture Assimilators for white or other dominant groups to persons who have some self-interest in good relationships with persons from other cultures, such as teachers, medical staff and people working in the domain of law.

Triandis (1975) realizes that conflicting interests between groups can be important factors in different perceptions and intergroup hostility. In the Culture Assimilator procedure conflicts are usually ignored. However, especially in the application of the Culture Assimilator for the purpose of improving interethnic relations within one country, it seems evident that these external factors have a critical effect upon the explanations given. An example from the Culture Assimilator developed by Landis *et al.* (1976) serves to illustrate this point. The situation concerns the promotion of army soldiers. A black soldier who has applied for a promotion visits his white Commanding Officer (CO) to ask why he has not been promoted. The soldier claimed that he had good scores and asked the CO to review his decision. After he has left, the white CO wonders why blacks in particular request promotion reviews. The 'correct' answer would be that 'blacks feel they will not be given a promotion unless they ask for one' (Landis *et al.*, 1976, p.175). One of the 'wrong' answers was that the CO was prejudiced and promoted more whites than blacks. This alternative was 'wrong', because discrimination could not be inferred from the description of the situation. However, if one knows that in general blacks do have fewer opportunities than whites, are being discriminated against and are judged differently because of existing stereotypes (cf. the chapter by Pettigrew & Martin in this volume), one cannot say that this answer is 'wrong', even if it is of another, 'higher' order than the 'correct' answer. One could argue that blacks themselves explain their behavior in this way, which makes it the 'correct' answer by definition. Apart from the problem of 'right' and 'wrong' in this context, one could say that even the 'correct' answer implies a disfavoring of blacks that makes it necessary for them to try harder than whites, at least in their own eyes. Because many contacts between blacks and whites, dominated and dominant, are characterized by discrimination, it is necessary to take that discrimination into account if one wishes to gain insight into the problems that may arise in interethnic contact. As long as Culture Assimilator training ignores the real conflicts and the existing discrimination between groups, there seems to us little reason to expect significant changes in attitudes and behavior from it. But this observation points to a more general problem to which we now turn – the emphasizing of similarities and differences between groups.

## THE PROS AND CONS OF EMPHASIZING DIFFERENCES

We have noted that most research on changing interethnic attitudes and behavior is based on the so-called Contact Hypothesis. Research based on this hypothesis has shown that, under particular conditions (for instance, the equal status of the participants), the attitude toward outgroup members changes. However, this change in attitude may not generalize to the outgroup as a whole. Hewstone and Brown (1986) account for this phenomenon by a change from an intergroup into an inter-individual relation during the contact situation. The fact that only the persons with whom the contact is established are evaluated positively can be explained by the 'ultimate attribution error' (Pettigrew, 1979). This expression denotes the tendency, based on negative stereotypes of outgroups, to attribute negative acts performed by outgroup members to enduring dispositions and to attribute positive acts to situational or transitory causes. In this way, positive behavior that does not fit in with the general evaluation and the stereotypes of the outgroup can easily be reasoned away and seen as exceptional. Therefore, Hewstone and Brown maintain that the attitude towards the outgroup as a whole can only be changed when the outgroup member with whom the contact is established is perceived as a typical outgroup member. In one of the few studies in which this hypothesis is tested, Weber and Crocker (1983) have demonstrated that their subjects' stereotypes were influenced more when the stereotyped person who did not behave stereotypically in a pleasant encounter was seen as a representative of the group, than when he was perceived as an atypical group-member. This view conflicts with that of Miller and Brewer (1984), who argue, in line with the Contact Hypothesis, that only emphasizing (experiencing) of (personal) similarities has positive effects on intergroup relations. They do not seem to acknowledge the existence of real differences between ethnic groups, for instance in religion, societal position and cultural norms and traditions. These general cultural differences also imply interpretational or attributional differences. The following example of interethnic contact in the classroom between a white Dutch teacher and a Turkish child serves to illustrate this point. When a Turkish child is reprimanded by the teacher, the child looks down. This behavior in Dutch society is interpreted as a sign of unreliability, disrespectfulness or humbleness, all negatively valued characteristics, but in the Turkish culture, looking down is a sign of respect. If one wants to change interethnic attitudes and behavior, these types of intercultural difference have to be taken into account by making them intelligible.

The Culture Assimilator provides one way to accomplish this. The expected changes brought about by the Culture Assimilator procedure will also influence the expectations the ingroup has about the behavior of outgroup members. Not only will it alter stereotypical expectations, but it also will 'build' new, realistic expectations. In chapter 2 of this volume Hewstone presents a model of the path by which interethnic conflict may be reduced through actual contact. In his model, initial negative expectations are assumed. Change should be brought about by a 'typical outgroup member' who disconfirms these negative expectations. Hewstone, however, points out the problem of making the ingroup members perceive the outgroup member as typical, because during the contact the intergroup situation may change into an inter-individual relation.

A second problem is that in many contact situations it is only realistic to expect differences in behavior. Therefore, it may be most fruitful to teach people to expect (and respect) these differences. Influencing expectations has another advantage over the creation of contact situations in which the outgroup member disconfirms expectations: it will be easier to carry out on a large scale (outside the laboratory). We therefore conclude that a good way to start reducing intergroup conflict is by changing expectations about the behavior of the outgroup members.

Although emphasizing differences seems necessary, some remarks must be made about possible side-effects. As has been amply demonstrated both empirically and theoretically (e.g., Tajfel, 1978), emphasizing, and therefore also perceiving, differences between the ingroup and the outgroup may exert a negative effect on intergroup feelings and behavior. One might therefore suppose that when someone has completed the Culture Assimilator training procedure, his or her general feeling and behavior might not be altered (or altered in the wrong direction), although the person is able to make isomorphic attributions and think less stereotypically. This may explain why expectations of the Culture Assimilator with regard to behavioral and affective effects have not been fulfilled. Of course, it is certainly possible that knowing more about the behavior of another group can lead to the positive experience of a greater feeling of control, without promoting a liking for the members of the other group. One may even expect that it gives pleasure to be able to manipulate others, which is made possible by going through a training procedure like the Culture Assimilator. An American businessman may ask at length about the Brazilian associate's health, wife and children knowing (from the Culture Assimilator) the positive effect of those questions on the interaction process, but feel at the same time a deep disdain for this need for self-disclosure. In other words, feelings (and behaviors) towards the members of the other culture are not necessarily affected in a positive way by knowing better how to interpret and anticipate their behavior.

The Culture Assimilator can in some situations provide more than one 'correct' explanation, instead of just one stereotype. It may thus create uncertainty about what to expect when one is confronted with a foreigner of whom one knows nothing. In this case, one has to learn to postpone judgement until more knowledge is gathered. Hence, it might be advisable to add other stories to the Culture Assimilator, in which the behavior of the outgroup is explained by them just as the ingroup explains it, but the ingroup interprets it differently when a member of the outgroup exhibits that same behavior (Sagar & Schofield, 1980). By doing so, similarities are also emphasized and people learn to expect not only differences, but also similarities which may decrease the 'us and them' feeling and have a more positive effect on interethnic attitudes and behavior (Stephan, 1985). The assumption that emphasizing similarities instead of emphasizing differences leads to liking is consistent with the findings in the psychology of emotions. Strangeness is discomforting: familiar stimuli are liked better than unfamiliar ones, just because they are familiar (Zajonc, 1968). However, strangeness can also give rise to curiosity (Allport, 1954). This raises the question in which conditions unfamiliarity is upsetting and in which it is interesting. Frijda (1986) supposes that this is dependent on the availability of coping reactions, that is, ways in which to deal with the situation. Interactions with strangers in general, and ethnic minorities in particular, may evoke a general feeling of anxiety because of their very strangeness, but if the (anticipated or imagined) interaction situations are perceived as controllable, this specific situation may evoke curiosity. Culture Assimilator training may increase the controllability of possible encounters with members of an ethnic minority group by teaching people what to expect and how to behave (See also Dijker's chapter in this volume on the relation between controllability and affect).

Apart from the general effects of emphasizing differences and possible 'inconsistencies' between thinking and feeling, one can differentiate between 'likeable' and 'unlikeable' differences. From the Culture Assimilator one learns where intergroup differences stem from, but this does not necessarily mean that one likes this other, 'different' group, nor that one appreciates these different behaviors. For instance, knowing that, within some Islamitic groups in the Netherlands, women are not allowed to go out without company, might become intelligible by knowing more about their cultural background. It does not mean, however, that one approves of these practices (which are seen as repressive by the dominant Dutch culture), or that one wants to interact with these 'dominant men' or 'their submissive women'. Acknowledging this, it seems more appropriate to work towards a change from general negative expectations to a more complex set of beliefs about the outgroup. It is not reasonable to expect a change from negative to absolutely positive expectations. A more appropriate goal would be to aim for mixed expectations.

The result will be more comparable to the stereotypes one holds about one's ingroup: although the general attitude is positive, people acknowledge characteristics of their own ethnic group which are not all valued positively. Just as having friends does not necessarily mean you like all their habits, one does not need to approve of all customs and values of another ethnic group to develop a positive attitude towards the group.

A third problem with emphasizing general differences is that often, within an ethnic group, many subgroups exist with their own quite different customs, values and characteristics. This means that, in many cases, it is simply not justified to speak of the 'typical' outgroup explanation for a given behavior. Just as with the aforementioned problem, the solution may be in trying to create a more 'complex view' of the outgroup: in some items one instead of two alternatives will be appropriate.

CONCLUSIONS

In sum, then, we see a future for the Culture Assimilator training procedure. It seems possible to improve interethnic relations by emphasizing ethnic differences, though probably not if that is the only thing being done. In order to attain this goal, not only the attributions the ingroup members make for the behavior of the outgroup, but also their attitudes and behaviors have to be changed. We suggest the following guide-lines for the construction of new Culture Assimilator training procedures.

1. *Trainees should have some personal or professional interest in the training procedure.* It seems advisable to develop assimilators for those groups of the dominant culture who have a self-interest in good interethnic contacts, who are (in part) dependent on them. In these cases, one could expect the greatest impact of the training procedure. This impact may even become stronger if the target group has a model function and/or social agent role in society for others (such as teachers).

2. *Besides differences, similarities between the groups should also be stressed.* To stress only differences can lead to more favorable cognitions, but does not necessarily promote consistent affective and

behavioral improvement. This result may occur because, when a group is perceived as being different, understanding these differences does not imply liking these differences. Thus, emphasizing similarities in addition to differences may make the ingroup see the outgroup in a more differentiated way: in some ways they are seen as 'us', and in others as 'them'.

*3. Besides differences between the groups, differences within the outgroup should also be emphasized.* One of the main problems in changing the stereotypes about an outgroup is that there is a tendency not to see any differences between the members of an outgroup. If one tries to substitute one 'stereotype' for another, this tendency may continue. It may, therefore, be more profitable to try to have people develop more complex views about the outgroup by teaching them to distinguish between different subgroups of the outgroup.

*4. The items should not only concern attributions about the behavior of members of the target culture but also attributions and behavioral alternatives of members of the own culture.* It is not only necessary to interpret correctly the behavior of members of the target culture, but also to know how one's own behavior is perceived and which behavioral alternatives are considered appropriate. Including such items in the Culture Assimilator may have a positive side effect: The emphasis is not totally on the 'different' behaviors of the other group and explanations thereof; the behavior of the own group is also open to question.

*5. The Assimilator should not ignore conflicts between the ingroup and the outgroup.* Many interethnic contacts can be characterized as discriminatory, or at least as influenced by power relations between dominant and dominated; this should be taken into account in a Culture Assimilator. Cultural differences may serve as a possible justification for disliking and discriminating against the outgroup. Hence, it is advisable not to play along with a tendency to reduce racist relations to cultural differences, or interethnic conflict to mistakes due to diverging interpretations.

We realize that these guidelines, in addition to the problems mentioned earlier about the construction of a Culture Assimilator, constitute a formidable task, but the effort seems worthwhile.

NOTE

* We are grateful to Thomas F. Pettigrew for his constructive comments on earlier drafts of this chapter.

# 13

# TOWARDS A USEFUL SOCIAL PSYCHOLOGY FOR ETHNIC MINORITIES

Jan Pieter van Oudenhoven
Groningen University
The Netherlands

Tineke M. Willemsen
University of Amsterdam
The Netherlands

The preceding chapters have described recent theoretical viewpoints on intergroup relations, presented several illustrations of interethnic behavior in real situations, and offered some strategies for prejudice reduction. In the next section we will briefly comment on these discussions, and evaluate their contribution to a useful social psychology for ethnic minorities. Subsequently we will discuss the effectiveness of theoretically based strategies to reduce intergroup conflicts, and the rather problematic relation between theoretical viewpoints and the applied social psychology of intergroup relations. In the concluding section of this chapter, on assimilation and pluralism, we will make a plea for the development of a more realistic and less pretentious program of research and application, in which we do not try to change deeply ingrained prejudices and behavior all at the same time, but aim instead to modify behavior toward a level of mutual tolerance.

THEORY, REAL LIFE, AND REMEDIES

*Theoretical developments*

Theories of prejudice have become increasingly cognitive during the last decade. Within this cognitive approach Tajfel's social identity theory, although definitely not restricted to cognitive phenomena, has undoubtedly been most influential. Tajfel's impact on the development of theories of intergroup relations is comparable to that of Allport and Sherif in previous decades. Not surprisingly, three of the four theoretical chapters in this book strongly reflect the influence of social identity theory.

Hewstone's chapter presents a fruitful integration of social identity theory and attribution theory. One of the practical implications of Hewstone's analysis of intergroup relations is that, in order to obtain generalization of positive attitudes, intergroup contact has to take place between typical members of the various groups, if they are not to be perceived as exceptions to their groups. This is a very important, but nonetheless controversial issue. Brewer and Miller (1984) and Turner (1981) for instance, contend that salience of group boundaries should be reduced to a minimum so that persons hardly realize that the other person is a typical member of his/her group, and that interaction should take place on a strictly interpersonal basis in order to improve interethnic relationships. As these two standpoints are not only interesting from a theoretical point of view, but lead to different strategies for the reduction of interethnic conflict, they are a research topic of primary importance.

Doise and Lorenzi-Cioldi's chapter provides a substantial contribution to and refinement of social categorization theory. In particular, Doise and Lorenzi-Cioldi dispute the axiom that intra- and intergroup differentiation are inversely related. They offer considerable empirical evidence that the correlation between ingroup solidarity and outgroup antagonism is not always strongly positive and sometimes even negative. Intergroup differentiation and, more specifically, intergroup hostility can generate intragroup differentiation and conflict as well. Doise and Lorenzi-Cioldi show that several well established theoretical models, explaining the patterns of differentiation within and between groups, are now available in social psychology. They suggest that researchers of intergroup relations, instead of arguing in favor of their own models, should try to integrate or articulate them and determine under which conditions different models apply. This is especially required when socially relevant issues are at stake, as is the case in interethnic relations.

Van Knippenberg has creatively worked out the strategies group members may resort to in their search for a positive social identity. He discusses the link between some socio-structural variables and strategies of identity management. Of particular interest is his focus on permeability of group boundaries--that is, the extent to which individuals are free to join or leave social groups--as a socio-structural variable. Societies tend to have stable intergroup relations when there is either no indiviual mobility across groups, as in a caste or feudal system, or where there exists a high level of permeability of group boundaries, as in a fairly open society. In both cases the intergroup relations are seen as legitimate. Relations between minority and majority groups both in Europe and the United States often lie somewhere between these two situations. They are characterized by a limited degree of intergroup permeability with instability as a consequence. There are various minority groups who are trying to develop positive ingroup identities by challenging the legitimacy of the existing status relationships. They may undertake collective actions to enhance positive ingroup identity (e.g. emancipation movements). Ethnic unrest should thus not be looked upon exclusively as a negative affair. Instability between ethnic groups may also be seen as a sign that boundaries between minority and majority are perceived as less impermeable than they used to be.

In his chapter 'Ethnic attitudes and emotions' Dijker points out the role of emotional aspects of ethnic attitudes, in particular emotional reactions for which persons themselves are not able to provide reasons. Knowing the determinants of these feelings may be crucial for understanding the elusiveness and perseverance of negative ethnic attitudes. Attitude is a dynamic concept. Most cognitive approaches to prejudice acknowledge this point but they typically emphasize subjects' activity and flexibility in the domain of knowledge. They neglect, however, an integral account of the regulation of thoughts, bodily changes associated with emotion, and expressive behavior. It may, for instance, be important to know what happens in thought when an attitude is not given voice because strong social norms prevent it from being expressed.

These theoretical developments are still geared more to explaining why prejudice and discrimination occur than to pointing out solutions. However, as we have seen, most of them do provide insight into the conditions necessary to promote a reduction of intergroup confict as well, e.g. Hewstone's attention to the typicality of a minority group member, and Van Knippenberg's taking into account the permeability of group boundaries. Just as the contact hypothesis has been refined over the years to include the specific conditions in which contact is useful for the diminishing of prejudice, so the cognitive theories on intergroup relations are growing into a more specific knowledge of the conditions in which the individual cognitive processes are likely to change into a less prejudiced direction.

*Real life studies*

In part II of this book four empirical studies on interethnic relations in concrete settings were discussed: peer relations at school, police practices, interethnic contacts in an urban neighborhood, and daily conversations. The general picture is moderately positive as far as the majority's overt behavior is concerned: hardly any signs of blunt discrimination were found. But with regard to the majority's cognitions and attitudes and the subtle ways in which they are expressed there seems to be less reason for optimism. Saharso's study shows that in spite of the fact that students from different ethnic backgrounds get along with each other quite well, friendships across ethnic borders are very rare and the attitudes of some of the majority students are prejudiced. Her chapter provides some clear examples of aversive racism (cf. Gaertner & Dovidio, 1986b), in which an apparent rejection of racial prejudice and discrimination is combined with an avoidance of students of other ethnic origins in friendship relations.

Bernasco and Van Schie find no evidence of any discriminatory application of jurisprudential rules by police officers toward ethnic juveniles. However, these ethnic juveniles do face a higher probability of being arrested. Perhaps, their greater chance of being arrested reflects a biased attitude of police officers.

De Jong describes how immigrants, by getting involved in the social life of a neighborhood, gradually become more accepted by the autochthonous residents. In the neighborhood a situation of ethnic tolerance develops: people of different ethnic backgrounds get along with each other satisfactorily. However, this does not imply that prejudices have ceased to exist.

Van Dijk demonstrates that daily conversations display many signs of prejudiced attitudes, although people tend to express their prejudices in a covert way. In many ways they do their best to present themselves as non-prejudiced ("I have nothing against foreigners, but...") but at the same time they cannot help expressing their biased opinions against ethnic groups.

An important feature of these empirical chapters is that they are not only concerned with abstract ideas like prejudice and stereotypes, but also depict in detail behavioral variables that show what is really going on in these situations. A slightly optimistic view about interethnic relationships can be derived from these studies: in certain situations, such as the police station – probably because of strong official anti-discrimination norms – or the neighborhood centre – where residents from different ethnic backgrounds share a common interest to achieve better living conditions – people seem to overcome their prejudices and at least behave in a non-discriminatory way. According to De Jong, it is neither realistic nor necessary to strive for an attitudinal change to accompany these behavioral changes: as long as people tolerate each other and each other's way of living, it is of minor importance whether, deep in their hearts, they really like each other.

*Remedies*

Traditionally, social psychological strategies for the reduction of interethnic conflicts focused on attitude change as a way to bring about unprejudiced behavior. Modern social psychology emphasizes that altered behavior is more often the precursor of altered attitudes (Aronson, 1969; Bem, 1972) rather than the other way around. Behavior, in turn, may be shaped to a considerable extent by the situation in which it occurs. This is the point of view from which Pettigrew and Martin (chapter 10) discuss several strategies for the inclusion of minority groups in organizations. According to these authors, social psychological remedies such as training or the creation of task interdependence may be useful attempts to alleviate prejudice and discrimination in organizations, but have to be used in conjunction with broader structural changes that alter the organization's personnel distribution. Their chapter is a state-of-the-art account of the way in which a combination of knowledge from research based on the contact hypothesis and from research on social cognition can be used for the implementation of minority participation in organizational settings.

Van Oudenhoven's chapter concentrates on cooperation, one aspect of the contact hypothesis, in the specific setting of the classroom. It fits in well with the modern social psychological approach, for cooperation may be considered a modification of behavior – evidently with support from school authorities – aimed at improving interethnic attitudes.

Van den Heuvel and Meertens' chapter on the Culture Assimilator deals with a method that was originally designed for Americans going abroad for work or study but is also currently being applied to improve black-white relations. Training with the Culture Assimilator generally results in a cognitive change, i.e. a better understanding of the outgroup, but attitudinal or behavioral changes toward the outgroup members are less clear, or even absent. The Culture Assimilator is not a very suitable instrument for large-scale use, since it is designed for use by specific groups such as people going abroad to work or – more relevant to the present context – for teachers working with ethnically heterogeneous groups. Van den Heuvel and Meertens' chapter again makes clear that programs that fail to take some behavior modification into account have a limited value.

Together, these chapters show that progress has been made in the way in which social psychologists are able to apply their knowledge to actual interethnic situations. However, at the same time, many pitfalls remain; more research is needed to establish optimal combinations of measures for different situations. Programs that combine behavioral measures with endeavors toward attitude change will problbaly prove to be the most effective ones. We will further comment upon the effectiveness of strategies in the next section.

## EFFECTIVENESS OF THEORETICALLY BASED STRATEGIES TO REDUCE INTERGROUP CONFLICT

Most theories on intergroup cognitions and behavior have been developed because their founders saw prejudice and discrimination flourishing around them, and wanted to understand these phenomena better in order to be able to counter them. However, not all theories have actually led to useful methods for the reduction of prejudice. We will now discuss the effects of the remedies that have been developed out of the main theoretical standpoints.

Many studies on changing intergroup behavior are based upon the contact hypothesis, which is, as we have seen, more of a set of hypotheses about the reduction of prejudice and the effects of interaction on attitudes than an integrative theory about the origins of prejudice. Actually, the contact hypothesis started as a set of conditions that had to be met in order for interracial attitude change to take place. Only afterwards has it acquired the status of a more general theoretical approach. On the whole, empirical evidence suggests that intergroup contact leads to a more favorable attitude of the prejudiced majority towards the minority, but only under certain conditions. Stephan (1985) catalogues the situational and personal variables that mediate the effect of intergroup contact. Of the situational variables, the effects of the nature of the interdependence between groups (i.e. cooperation or competition) have been specifically studied. Generally, cooperation leads to more favorable attitudes and relations between groups than competition. In order to achieve favorable intergroup attitudes as a result of cooperative intergroup contact it is important that a positive outcome is experienced. Equal status of the members of the interacting groups is another important condition.

The contact hypothesis, as it was originally formulated, is an example of the situational approach: the emphasis is not on the prejudiced individual, but on the circumstances that allow the individual's prejudice to be changed. Educational approaches using information or propaganda have the same emphasis. The aim of such programs is to make the subjects' 'informational environment' less biased against minority groups. Generally, strategies based on transfer of information are not very effective.

Remedy actions based on psychodynamic (symptom) theories are focused on individuals or small groups. Individual treatment can never have a wide social impact. Individual therapy may well be helpful in some cases, but can, of course, never offer a solution to such a large scale phenomenon as ethnic prejudice. Small group therapies, discussion groups and workshops often show positive results, although some precautions have to be made, because the possibility exists that persons with strong prejudices end up with even more extreme attitudes (cf. Myers & Bishop, 1970). This tendency of individuals with extreme prejudices to become even more prejudiced when participating in discussion groups, may only be a specific case of a well-known phenomenon, the so-called group-polarization of attitudes. The Culture Assimilator may also be considered an individual or small group strategy to reduce prejudices, but, as we have seen above, its effects are primarily confined to a better understanding of the outgroup.

Social categorization theory as well as other cognitive theories have not produced many practical remedies. Billig (1985) has criticized social categorization theory and other cognitive theories for that very reason: by explaining cognitive biases, such as stereotypes, as functional forms of cognitive efficiency, the cognitive perspective becomes less amenable to intervention and so confirms the status quo, which is one of discrimination and racism. Tajfel (1978, pp.93-98) proposes basically two general strategies for minority groups to enhance their social identity. One tactic consists of assimilation of the minority group into the dominant group's power structure. In many cases this would first require some adaptations in the legislation. However, assimilation always implies adapting oneself to what the majority has already thought up – a majority, moreover, which will not be very willing to change the laws. Where, for whatever reasons, the minority group or its individual members are not able to assimilate into the majority group, another strategy to acquire a positive social identity becomes necessary. The most likely strategy consists of the re-evaluation of dimensions or concepts which negatively influence self-identity (the 'Black is beautiful' approach) or to create new characteristics which have a positively valued distinctiveness from the majority group. This strategy may be helpful for the enhancement of the ingroup-identity, but does not necessarily lead to less prejudice and more acceptance by the outgroup; without acceptance of the re-evaluated or new characteristics by the majority the strategy will not be successful.

Pettigrew and Martin (in chapter 10 of this book) have succeeded in proposing a coherent set of practical measures which are based on a combination of cognitive and situational theories of prejudice. One of their major premises, derived from the situational approach, is that modifying behavior by shaping situations is more successful than attempts to change attitudes. At the same time they explicitly take into account such cognitivist phenomena as the extreme evaluation of persons in solo positions (a solo is a single minority group member in a majority group) when suggesting ways to diminish discrimination in organizations.

PROBLEMS CONCERNING THE RELATION BETWEEN THEORIES AND APPLICATION

Frequently, theories offer only few suggestions to reduce prejudice, whereas practical solutions often lack a theoretical foundation or even contradict the dominant theoretical trends. One possible explanation for this mismatch is that theoreticians and practicioners have different outlooks. Theoreticians try to understand behavior; their purpose is to establish rules of behavior. Since prejudice is a very general and also a very persistent phenomenon, unprejudiced behavior will be seen as an exception to the rule and consequently get less attention. Practicioners, on the other hand, are pursuing changes in behavior, which implies that they try to provoke exceptions to the rule, i.e. unprejudiced behavior. Thus one might say that theoreticians tend to be somewhat conservative because they like to see their models of behavior confirmed whereas practicioners prefer to see exceptions to these models since they are committed to change. Moreover, practicioners are more interested in actual behavior, whereas the current mainstream theoreticians focus on the mechanisms underlying this behavior, such as cognitions which are generally more resistant to change.

Many remedy programs that are focused on attitude change seem to ignore that the relation between cognition and behavior is not at all linear and sometimes even absent. Educational programs in particular too easily assume that changing a cognition will lead to less prejudiced behavior. However, as we have pointed out in the introductory chapter, there is little evidence for a strong correlation between individual cognitions and behavior as far as interethnic relations are concerned. Situational aspects, such as group norms, are somewhat more influential. Of course, for psychologists, it is always tempting to pay attention to cognitions; but social psychologists who want to improve interethnic relations might better concentrate their efforts on the social context, the situational aspects.

A consequence of the current cognitive trend in social psychology is that little attention is paid to larger societal structures. The emphasis on individual cognitive processes and their influence on people's behavior has made it easy to overlook such larger structures. In fact, they are often not considered part of social psychology, but rather seen as belonging to the sociologists' frame of reference. Yet, it is evident that macro-aspects such as socio-economic differences or legal structures do play an important role in interethnic relations and attitudes in addition to the individual cognitive and behavioral processes social psychologists usually study. Pettigrew and Martin's analysis of organizational inclusion of minority groups (in this book) is one of the few social psychological approaches that does take these larger societal structures into account.

Another drawback of the cognitive trend in the social psychology of intergroup relations is that affective considerations are virtually ignored. This is surprising considering the emotional character of many interethnic conflicts. Social psychologists may well be able to explain interethnic conflicts but they have not yet succeeded in developing adequate scenarios to cope with sudden outbursts of ethnic unrest, which are often highly emotional events.

Finally, to be able to measure improvement in interethnic relations, it is crucial to choose the appropriate dependent measure. If we are trying to improve behavior, it might be more convenient to attempt to actually measure behavioral variables like the behavioral tolerance suggested by De Jong (Chapter 8), instead of refining our measures of prejudice. We are able to demonstrate the existence of prejudices but, unfortunately, it hardly helps us to explain or predict (un)prejudiced behavior. This is not a plea to bury one's head in the sand and act as if prejudice does not exist; it is a plea, instead, to concentrate our efforts, both in theoretical and in applied studies, on the behavioral aspects we most urgently want to modify. Moreover, in this way it will be easier to keep believing in the possibilities of indeed improving interethnic relations, and not to fall victim to the pessimism that, according to some authors, is inherent in cognitive theories of prejudice (cf. Billig, 1985).

ASSIMILATION VERSUS PLURALISM

Assimilation and pluralism form the two poles of a dimension reflecting the degree to which a minority group is supposed to adapt to the majority. Theories and strategies in the field of inter-group relations may be ordered according to the degree to which they adopt an assimilationist or pluralistic point of view. For instance, from the chapters in this book, Pettigrew and Martin generally present assimilationist strategies (cf. Feagin, 1987), while De Jong emphasizes pluralism, a tolerance for the existence of different ethnic groups within one neighborhood who explicitly present themselves as distinct groups. Social categorization theory clearly implies an assimilationist approach through its assumption that the reduction of the salience of group boundaries will improve intergroup relations. And Deschamps and Doise's (1978) suggestion of cross-cutting categorical distinctions, for instance, through inter-marriage of members of different ethnic groups, is even a quite explicit kind of assimilation.

Assimilation and pluralism are not only theoretical terms but, more importanty, constitute an ideological issue as well. For a long time assimilation has been the predominant ideology, particularly in the United States: the melting pot ideology, a fusion of various elements into a new homogenised whole. In practice, this has to a major extent resulted in an adaptation of all other groups to the predominant Anglo-Saxon culture, with the remains of their original languages and cultures becoming more folklore than a cultural tradition. Similar tendencies to assimilate minority members can be observed in most Western European countries where new ethnic minorities (e.g. refugees from the Far East, immigrants from former colonies, or immigrant workers from Mediterranean countries) have been arriving during the last decades. In most countries with ethnic minority groups, however, there exist firm social barriers between the various groups; in many cases the minority members are reluctant to give up their cultural and social identities and do not wish to assimilate into the larger culture.

In other words, not only because of the social barriers to becoming part of the majority group, but also because of a positive appreciation of their own culture, ethnic groups often adhere to cultural pluralism, i.e. the maintenance of their exclusive cultural inheritance within the larger system. Pluralism can take many forms; some groups stress the importance of a certain degree of adaptation to the larger system while others strive for maximal cultural autonomy. In our opinion, some form of pluralism is to be preferred over complete assimilation. One of the negative consequences of full assimilation is that a cultural vacuum among minority group members may develop. The second generation of immigrants in particular may lose their ethnic, linguistic or religious roots while not being adequately rooted in the majority culture either. This loss of identity and historical roots – being caught in between several cultural identities – may result in feelings of social alienation or marginality (Amir, 1984). In some cases (relative) cultural autonomy may be a realistic solution, especially when a cultural minority group is concentrated in a particular geographic area and has a socio-economic status comparable to that of the majority. One of the most successful examples of the existence of several relatively distinct cultural groups is Switzerland where four linguistic groups each have a separate but more or less equal status. However, the minority groups we have been referring to in this book are widely dispersed – in varying concentrations – throughout a country. Under these circumstances some form of modified pluralism (Glazer & Moynihan, 1963) may be a viable option. This option implies that groups may retain distinctive features although not necessarily those they originally possessed, since it is inevitable that groups change over time and are affected by the larger system.

## Cultura franca

Maintaining cultural diversity in combination with mutual respect between the different groups is an optimal situation which is rarely achieved. It is difficult to establish an acceptable form of pluralism. Too often, pluralism means a rigid adherence to one's own culture leading to a sharpening of intergroup discrepancies and to separatism or even interethnic conflicts, particularly when the different ethnic groups greatly vary culturally and have a history of negative relationships. In these cases – which are not at all exceptional – it may help to develop a common body of culture which does not belong to the cultural heritage of either group or which is a transformation of an aspect of the culture of one of the groups. We will call such a body of culture 'cultura franca', using the analogy of lingua franca. A lingua franca is a language which is used as a means of communication among people who have no language in common. It may not even have native speakers, as for instance Swahili in some East-African countries or Latin in medieval Europe.

Language usually has two important functions, a communicative and a symbolic one; in addition to being a tool of communication it is an emblem of groupness as well (Edwards, 1985). The interesting characteristic of a lingua franca is that its primary function – communication – takes place across group boundaries. By using a language that does not evoke the specific emotional evaluations that group-tied languages do, the salience of group membership may be reduced; at the same time groups that participate in the communication do not have to discard their own language altogether. If they had to discard their own language, other differences between the groups would possibly become more salient in order to maintain their social identity. Eventually the lingua franca may even become the superordinate emblem of two or more linguistically distinct groups (Swahili has to a certain degree become a superordinate language for several tribes in Eastern Africa, even across the national borders). Similarly, a cultura franca may provide an area of communication in which majority and minority members participate on a common basis without experiencing ethnic sentiments. A lingua franca may constitute a major element of the cultura franca. Of which aspects may a cultura franca further consist? It is very difficult to make general statements since it will depend on the specific cultures of the participating groups and the kind of communications in which they want to get involved. It may be founded upon a mutual respect for each others' religion, or there may be shared norms about how trading should take place. Often sports offer a topic of common interest, although sports may reinforce ethnic conflict if competition takes place between ethnic groups. On a local level a cultura franca may develop when several ethnic groups work for a common interest, such as the improvement of their neighborhood, the acquisition of sport facilities etc. De Jong (chapter 8 of this book) presents a description of such a process on the neighborhood level. Generally, a cultura franca may develop out of a necessity to communicate for practical purposes; emotionally laden issues can be avoided. What we are proposing here is to broaden the base of cultura franca by deliberately promoting more common interest activities that are free from group tied emotions.

The idea that the formation of a common culture may provide a means to promote integration is supported by experiences with mergers of organizations, which also often involve emotional conflicts because of diverging cultures. According to Gaertner and Dovidio (1986a) successful mergers are associated with an integration pattern in which in the first stage the two management teams join together without either company being required to conform to the style of the other, followed by a second stage in which the merged company develops a culture that represents a blend of the two corporate cultures or combines each into a pattern in which new values, norms, and procedures emerge. There may even arise a completely new culture. We do not suggest that ethnic groups should gradually lose their identities (even merged oranizations may keep a great deal of their original identities), but some form of voluntary integration is desirable in order to be able to communicate adequately.

Communicating without too much engagement, as it takes place within the domain of a cultura franca, is in accordance with the way people are willing to behave. Schofield (1986), for instance, states that blacks and whites are more willing to engage in fairly impersonal interactions than in close personal ones. Moreover, the participants do not have to behave in a forced 'color-blind' way according to the taboo that seems to exist in our current western society against making overt references to race. In a society where race and skin color are salient perceptual features, such insistence on denial of racial differences endangers the learning of the critical lesson that race is basically an irrelevant social category (Schofield, 1982).

CONCLUSION

The chapters in this book have shown that social psychology may make contributions that are both theoretically and practically relevant to a major societal problem. However, it is also clear that social psychologists can seldom report important reductions of prejudice or of discriminatory behavior. Therefore, social psychologists should realize that interethnic prejudice is such a large scale phenomenon that it would be pretentious to strive for improvements of any magnitude while neglecting larger societal structures such as the legislature or organizational recruitment and selection procedures.

For the time being it is, in our view, advisable to concentrate our efforts on what has proven to be most successful: the reduction of discriminatory behavior on individual, small-group, and organizational levels. Changing individuals' prejudices seems to be a very hard task; if it is possible at all, we find that in many instances the modified attitude is not followed by a change in behavior. If, however, we start by changing behavior, there is a good chance that changes in prejudice will follow. Moreover, from a practical point of view changing behavior is more urgent than changing cognitions, especially in situations of ethnic conflict or discrimination against ethnic minorities. It is in line with this focus on behavior that we suggest the development of a cultura franca for problematic interethnic relations. Different ethnic groups within one society always need to communicate somehow. By stressing contents and ways of communication that are relatively free of group-tied emotions a vehicle is created to fulfill the need for communication. The combination of solving practical communicative needs with the avoidance of emotionally laden topics may gradually reinforce the delicate process of a wider mutual understanding and acceptance.

ACKNOWLEDGEMENTS

The authors would like to thank Naomi Ellemers and Bert Wiersema for their critical comments on an earlier version of this chapter.

# BIBLIOGRAPHY

ABELSON, R. P. (1976). Script processing in attitude formation and decision making. In J. S. Carroll & J. W. Payne (Eds.), *Cognition and social behavior* (pp. 33-46). Hillsdale, NJ: Erlbaum.

ADORNO, T. W., FRENKEL-BRUNSWIK, E., LEVINSON, D. J., & SANFORD, R. N. (1950). *The authoritarian personality*. New York: Harper.

AJZEN,I., & FISHBEIN, M. (1977). Attitude behavior relations: A theoretical analysis and review of empirical research. *Psychological Bulletin, 84*, 888-918.

ALBERT, R. D. (1983). The intercultural sensitizer. In D. Landis & R. W. Brislin (Eds.), *Handbook of intercultural training: Volume II issues in training methodology* (pp.186-217). New York: Pergamon.

ALLPORT, G. W. (1948). *ABC's of scapegoating*. Freedom Pamphlets, Anti Defamation League of B'nai B'rith, New York.

ALLPORT, G. W. (1954). *The nature of prejudice*. Cambridge, MA: Addison Wesley.

AMIR, Y. (1969). Contact hypothesis in ethnic relations. *Psychological Bulletin, 71*, 319-342.

AMIR, Y. (1976). The role of intergroup contact in change of prejudice and ethnic relations. In P. A. Katz (Ed.), *Towards the elimination of racism*. New York: Pergamon Press.

AMIR, Y., SHARAN, S., & BEN-ARI, R. (1984). Why integration? In Y. Amir & S. Sharan, *School desegregation*. Hillsdale: Lawrence Erlbaum.

ARCURI L. (1982). Three patterns of social categorization in attribution memory. *European Journal of Social Psychology, 12*, 271-282.

ARONSON, E. (1969). The theory of cognitive dissonance: A current perspective. In L. Berkowitz (Ed.), *Advances in experimental social psychology*. Vol. 4. New York: Academic Press.

ARONSON, E., STEPHAN, C., SIKES, J., BLANEY, N., & SNAPP, M. (1978). *The jigsaw classroom*. Beverly Hills, CA: Sage.

ASHMORE, R. D. (1970). The problem of intergroup prejudice. In B. Collins, *Social psychology* (pp. 245-296). Reading, MA: Addison-Wesley.

ASHMORE, R. D., & DEL BOCA, F. K. (1976). Psychological approaches to understanding intergroup conflicts. In P. A. Katz (Ed.), *Towards the elimination of racism* (pp. 73-123). New York: Pergamon.

ASHMORE, R. D., & DEL BOCA, F. K (1981). Conceptual approaches to stereotypes and stereotyping. In D. L. Hamilton (Ed.), *Cognitive processes in stereotyping and intergroup behaviour* (pp. 1-35). Hillsdale, NJ: Erlbaum.

AUGUSTIN, V., & BERGER, H. (1984). *Einwanderung und Alltagskultur* [Immigration and every-day culture]. Berlin: Publica.

BARENDREGT, J. T., & FRIJDA, N. (1982). Cognitive aspects of anxiety. *Journal of Drug Research, 4*, 17-24.

BEALS, A. R. (1962). Pervasive factionalism in a south Indian village. In M. Sherif (Ed.), *Intergroup relations and leadership* (pp. 247-266). New York: Wiley and Sons.

BELL, D., & LANG, K. (1985). The intake dispositions of juvenile offenders. *Journal of Research in Crime and Delinquency, 22*, 287-308.

BEM, D. J. (1972). Self-perception theory. In L. Berkowitz (Ed.), *Advances in experimental social psychology*. Vol. 6. New York: Academic Press.

BERGER, J., WAGNER, D. G, & ZELDITCH, M. (1985). Introduction: Expectation tates theory: review and assessment. In J. Berger & M. Zelditch (Eds.), *Status, rewards and influence*. San Francisco: Jossey Bass Publishers.

BERLEW, D. E., & HALL, D. T. (1971). Socialization of managers: Effects of expectations on performance. In D. A. Kolb, I. M. Rubin, & J. M. McIntyre (Eds.) *Organizational psychology: A book of readings*. (3rd Ed.) Englewood Cliffs, NJ: Prentice-Hall.

BERNARD, J. (1981). *The female world*. New York: Free Press.

BERNASCO, W. (1987). *Autochtone en allochtone jongeren bij de jeugdpolitie: etnische afkomst, ongewenst gedrag en het optreden van de jeugdpolitie* [Native and non-native juveniles and the juvenile police; ethnic origin, undesirable behavior and police practices]. Internal report. Dept. of Social and Organizational Psychology, State University of Leiden (the Netherlands).

BERSHEID, E., & WALSTER, E. H. (1978). *Interpersonal attraction*. Reading, MA: Addison-Wesley.

BETTELHEIM, B., & JANOWITZ, M. (1950). *Dynamics of prejudice: A psychological and sociological study of veterans*. New York: Harper.

BILLIG, M. (1976). *Social psychology and intergroup relations*. European Monographs in Social Psychology. London: Academic Press.

BILLIG, M. (1985). Prejudice, categorization and particularization: from a perceptual to a rhetorical approach. *European Journal of Social Psychology, 15*, 79-103.

BILLIG, M. (1988). The notion of 'prejudice': Some rhetorical and ideological aspects. In T. A. van Dijk & R. Wodak (Eds.), *Discourse, racism and social psychology*. Special issue of Text 8, nr. 1, 91-110.

BILLIG, M., & TAJFEL, H. (1973). Social categorization and similarity in intergroup behaviour. *European Journal of Social Psychology, 3*, 27-52.

BIZMAN, A., & AMIR, Y. (1984). Integration and attitudes. In Y. Amir & S. Sharan, *School desegregation*. Hillsdale: Lawrence Erlbaum.

BLACK, D. (1971). The social organization of arrest. *Stanford Law Review, 23*.

BLACK, D., & REISS, A. J. (1970). Police control of juveniles. *American Sociological Review, 35*, 63-77.

BLALOCK, H. M. (1967). *Toward a theory of minority group relations*. New York: J.Wiley.

BODENHAUSEN, G. V., & LICHTENSTEIN, M. (1987). Social stereotypes and information-processing strategies: the impact of task complexity. *Journal of Personality and Social Psychology, 50*, 871-880.

BODENHAUSEN, G. V., & WYER JR, R. S. (1985). Effects of stereotypes on decision making and information-processing strategies. *Journal of Personality and Social Psychology, 48*, 267-282.

BOLTANSKI J. L. (1982). *Les cadres, la formation d'un groupe social*. Paris: Minuit.

BONACICH, E. (1972). A theory of middleman minorities. *American Sociological Review, 38*, 583-594.

BOND, M. H., HEWSTONE, M., WAN, K. -C., & CHIU, C. -K. (1985). Group-serving attributions across intergroup contexts: Cultural differences in the explanation of sex-typed behaviours. *European Journal of Social Psychology, 15*, 435-451.

BOVENKERK, F., BRUIN, K., BRUNT, L., & WOUTERS, H. (1985). *Vreemd volk, gemengde gevoelens* [Foreign people, mixed feelings]. Meppel: Boom

BOWER, G. H. (1981). Mood and Memory. *American Psychologist, 36*, 129-158.

BRADLEY, C. W. (1978). Self-serving biases in the attribution process: A re-examination of the fact or fiction question. *Journal of Personality and Social Psychology, 35*, 56-71.

BRAKE, M. (1985). *Comparative youth culture: the sociology of youth cultures and subcultures in America*. London: Routledge & Kegan Paul.

BRANTHWAITE, A., & JONES, J. E. (1975). Fairness and discrimination: English versus Welsh. *European Journal of Social Psychology, 5*, 323-338.

BREWER, M. B., & MILLER, N. (1984). Beyond the contact hypothesis: Theoretical perspectives on desegregation. In N. Miller & M. B. Brewer (Eds.), *Groups in contact: The psychology of desegregation*. New York: Academic Press.

BRISLIN, R. W., CUSHNER, K., CHERRIE, C., & YONG, M. (1986). *Intercultural interactions: A practical guide*. Beverly Hills: Sage.

BRONSON, G. W., & PANKEY, W. B. (1977). On the distinction between fear and wariness. *Child Development, 48*, 1167-1183.

BROWN, B. B., & LOHR, M. J. (1987). Peer-group affiliation and adolescent self-esteem: An integration of ego-identity and symbolic-interaction theories. *Journal of Personality and Social Psychology, 52*, 47-55.

BROWN, J. S. (1948). Gradients of approach and avoidance responses and their relation to motivation. *Journal of Comparative and Physiological Psychology, 41*, 450-65.

BROWN R. J., & ROSS G. F. (1982). The battle for accceptance: An investigation into the dynamics of intergroup behaviour. In H. Tajfel (Ed.), *Social identity and intergroup relations* (pp. 155-178). Cambridge: Cambridge University Press.

BROWN R. J., & TURNER J. C. (1979). The criss-cross categorization effect in intergroup discrimination. *British Journal of Social and Clinical Psychology, 18,* 371-383.

BROWN, R.J., & WADE, G. (1987). Superordinate goals and intergroup behaviour: the effect of role ambiguity and status on intergroup attitudes and task performance. *European Journal of Social Psychology, 17,* 131-142.

BROWN R. J., & WILLIAMS J. (1984). Group identification: the same thing to all people? *Human Relations, 37,* 547-564.

BYRNE, D., & WONG, T. J. (1968). Racial prejudice, interpersonal attraction and assumed dissimilarity of attitudes, *Journal of Abnormal and Social Psychology, 65,* 246-253.

CAMPBELL D. T. (1956). Enhancement of contrast as composite habit. *The Journal of Abnormal and Social Psychology, 53,* 350-355.

CAMPBELL D. T. (1967). Stereotypes and the perception of group differences. *American Psychologist, 22,* 817-829.

CANTOR, N., & MISCHEL, W, (1979). Prototypes in person perception. In L. Berkowitz (Ed.), *Advances in experimental social psychology.* Vol. 12. New York: Academic Press.

CAPLAN, N., & PAIGE, J. (1968). A study of ghetto rioters. *Scientific American, 219,* 15-21.

CHAIKEN, A. L., & COOPER, J. (1973). Evaluation as a function of correspondence and hedonic relevance. *Journal of Experimental Social Psychology, 9,* 257-264.

CHANCE, N. A. (1962). Factionalism as a process of social and cultural change. In M. Sherif (Ed.), *Intergroup relations and leadership* (pp. 267-273). New York: Wiley and Sons, .

CICOUREL, A. V. (1973). *Cognitive sociology.* Harmondsworth: Penguin.

CLARK, K. B. (1953). Desegregation: An appraisal of the evidence. *Journal of Social Issues, 9,* 2-76.

CLARK, K. B., & CLARK, M. P. (1947). Racial identification and preference in negro children. In T. M. Newcomb & E. L. Hartley (Eds.), *Readings in social psychology.* New York: Holt.

COHEN, A. R. (1958). Upward communication in experimentally created hierarchies. *Human Relations, 11,* 41-53.

COHEN, C. E. (1981). Person categories and social perception: Testing some boundaries of the processing effects of prior knowledge. *Journal of Personality and Social Psychology, 40,* 441-452.

COHEN, E. (1972). Interracial interaction disability. *Human Relations, 25,* 9-24

COHEN, E. G. (1980). Design and redesign of the desegregated school: problem of the status, power, and conflict. In W. G. Stephan and J. R. Feagin (Eds.), *School desegregation: Past, present, and future.* New York: Plenum.

COLLINS, S. M. (1983). The making of the black middle class. *Social Problems, 30,* 369-82.

COOK., S. W. (1969). Motives in a conceptual analysis of attitude-related behavior. In W. J. Arnold and D. Levine (Eds.), *Nebraska symposium on motivation.* Lincoln, NE: University of Nebraska Press.

COOK., S. W. (1978). Interpersonal and attitudinal outcomes in cooperating interracial groups. *Journal of Research and Development in Education, 12,* 97-113.

COOK., S. W. (1984). Cooperative interaction in multiethnic contexts. In N. Miller & M. B. Brewer (Eds.), *Groups in contact: The psychology of desegregation.* New York: Academic Press.

COOK, T. D., CROSBY, F., & HENNIGAN, K. M. (1977). The construct validity of relative deprivation. In J. M. Suls & R. L. Miller (Eds.), *Social comparison processes* (pp. 307-333). Washington DC: Hemisphere.

COOPER, J., & FAZIO, R. H. (1979). The formation and persistence of attitudes that support intergroup conflict. In W. G. Austin and S. Worchel (Eds.), *The social psychology of intergroup relations.* Monterey, CA: Brooks/Cole.

COX, O. C. (1948). *Caste, class, and race.* New York: Doubleday.

CROCKER, J., & MCGRAW, K. M. (1982). What's good for the goose is not good for the gander: Solo status as an obstacle to occupational achievement for males and females. *American Behavioral Scientist, 27,* 357-369.

CROCKER, J., HANNAH, D. B., & WEBER, R. (1983). Person memory and causal attributions. *Journal of Personality and Social Psychology, 44,* 55-66.

CROSBY, F. (1976). A model of egoistical relative deprivation. *Psychological Review, 83,* 85-113.

CROSBY, F., BROMLEY, S., & SAXE, L. (1980). Recent unobtrusive studies of black and white discrimination and prejudice: A literature review. *Psychological Bulletin, 87,* 546-563.

CROSBY, F. J. (1982). *Relative deprivation and working women.* New York: Oxford University Press.

DAVIS, G., & WATSON, G. (1982). *Black life in corporate America: Swimming in the mainstream.* Garden City, NY: Anchor Press/Doubleday.

DAVIS, J. A. (1959). A formal interpretation of the theory of relative deprivation. *Sociometry, 22,* 280-296.

DEAUX, K. (1976). Sex: A perspective on the attribution process. In J. H. Harvey, W. J. Ickes, and R. F. Kidd (Eds.), *New directions in attribution research* (Vol. 1). Hillsdale, NJ: Erlbaum.

DEAUX, K. (1984). From individual differences to social categories: Analysis of a decade's research on gender. *American Psychologist, 35,* 867-881.

DEAUX, K., & EMSWILLER, T. (1974). Explanations of successful performance on sex-linked tasks: What is skill for the male is luck for the female. *Journal of Personality and Social Psychology, 29,* 80-85.

DEAUX, K., WINTON, W., CROWLEY, M., & LEWIS, L. (1985). Level of categorization and content of gender stereotypes. *Social Cognition, 3,* 145-167.

DEN UYL, R., CHOENNI, C., & BOVENKERK, F. (1986). *"Mag het ook een buitenlander wezen?" Discriminatie bij uitzendbureaus* ["Is a foreigner also acceptable?" Discrimination by temporary employment agencies]. Utrecht: Landelijk Buro Racismebestrijding.

DEPREZ, K., & PERSOONS, Y. (1984). On the ethnolinguistic identity of flemish high school students in Brussels. *Journal of Language and Social Psychology, 3,* 273-296.

DESCHAMPS J. C. (1977). Effect of crossing category memberships on quantitative judgement. *European Journal of Social Psychology, 7,* 517-521.

DESCHAMPS J. C. (1982). *Social identity and relations of power between groups.* In H. Tajfel (Ed.), Social identity and intergroup relations. Cambridge: Cambridge University Press.

DESCHAMPS J. C. (1984). Identité sociale et différentiations catégorielles. *Cahiers de Psychologie Cognitive, 4,* 449-474.

DESCHAMPS J. C., & DOISE, W. (1978). Crossed category memberships in intergroup relations. In H. Tajfel (Ed.), *Differentiation between social groups* (pp. 141-158). London: Academic Press.

DESCHAMPS J. C., & LORENZI-CIOLDI F. (1981). "Egocentrisme" et "Sociocentrisme" dans les relations entre groupes [Egocentrism and Sociocentrism in intergroup relations]. *Revue Suisse de Psychologie, 40,* 108-131.

DEVRIES, D., EDWARDS, K., & SLAVIN, R. (1978). Biracial learning teams and race relations in the classroom: Four field experiments using teams-games-tournaments. *Journal of Educational Psychology, 12,* 28-38.

DIAB L. N. (1970). A study of intragroup and intergroup relations among experimentally produced small groups. *Genetic Psychology Monographs, 82,* 49-82.

DIJKER, A. J. (1987a). *Stereotyped-based information processing and cognitive load: The role of time pressure and accountability.* Unpublished manuscript, Psychological Laboratory, University of Amsterdam.

DIJKER, A. J. (1987b). Emotional reactions to ethnic minorities. *European Journal of Social Psychology, 17,* 305-325.

DIJKER, A. J. (1988a). *Emotional aspects of stereotypes.* Manuscript submitted for publication.

DIJKER, A. J. (1988b). *Mood effects on ethnic stereotyping.* Manuscript submitted for publication.

DIJKER, A. J. (1988c). Unpublished data. Psychological Laboratory, University of Amsterdam.

DIK, S. C. (1978). *Functional grammar.* Amsterdam: North Holland.

DION, K. L., & EARN, B. M. (1975). The phenomenology of being a target of prejudice. *Journal of Personality and Social Psychology, 32,* 944-950.

DION, K. L., EARN, B. M., & YEE, P. H. N. (1978). The experience of being a victim of prejudice: An experimental approach. *International Journal of Psychology, 13,* 197-214.

DOISE, W. (1969). Intergroup relations and polarizations of individual and collective judgments. *Journal of Personality and Social Psychology, 12,* 136-143.

DOISE, W. (1978). *Groups and Individuals. Explanations in social psychology.* Cambridge: Cambridge University Press.

DOISE, W. (1986). *Levels of explanation in social psychology.* Cambridge: Cambridge University Press.

DOISE, W., DESCHAMPS, J. C., & MEYER, G. (1978). The accentuation of intra-category similarities. In H. Tajfel (Ed.), *Differentiation between social groups* (pp. 159-168). London: Academic Press.

DOISE, W., & LORENZI-CIOLDI, F. (1987). *L'identité comme représentation sociale* [Identity as social representation], Paper presented at the Symposium "Représentations Sociales et Idéologies", University of Paris X-Nanterre.

DOISE, W., & SINCLAIR, A. (1973). The categorization process in intergroup relations. *European Journal of Social Psychology, 3,* 145-157.

DOLLARD, J., DOOB, L. W., MILLER, N. E., MOWRER, O. H., & SEARS, R. R. (1939). *Frustration and aggression.* New Haven: Yale Univ. Press.

DONNERSTEIN, E., & DONNERSTEIN, M. (1976). Research in the control of interracial aggression. In R. G. Geen and E. C. O'Neal (Eds.), *Perspectives on aggression.* New York: Academic Press.

DOOB, A. L., & CHAN, J. B. L. (1982). Factors affecting police decisions to takejuveniles to court. *Canadian Journal of Criminology, 24*, 25-37.

DOVIDIO, J. F., & GAERTNER, S. L. (Eds.). (1986). *Prejudice, discrimination, and racism.* Orlando, FL: Academic Press.

DUNCAN, B. L. (1976). Differential social perception and attribution of intergroup violence: Testing the lower limits of stereotyping of blacks. *Journal of Personality and Social Psychology, 34,* 590-598.

EDWARDS, J. (1985). *Language, society and identity.* Oxford: Basil Blackwell.

EISER J. R. (1971). Enhancement of contrast in the absolute judgment of attitude statements. *Journal of Personality and Social Psychology, 17,* 1-10.

EISER J. R., & VAN DER PLIGT, J. V. (1984). Accentuation theory, polarization and the judgment of attitude statements. In J. R. Reiser, (Ed.), *Attitudinal judgment,* New-York: Springer.

ELLEMERS, N., VAN KNIPPENBERG, A., DE VRIES, N., & WILKE, H. (1987). Antecedenten van tevredenheid en identificatie met de groep [Antecedents of satisfaction and identification with groups]. In A. van Knippenberg, M. Poppe, J. Extra, G. J. Kok & E. Seydel (Eds.), *Fundamentele sociale psychologie,* deel 2. Tilburg: Tilburg University Press.

ELWERT, G. (1982). Probleme der Auslander Integration [Problems of foreign intergration]. *Kölner Zeitschrift für Soziologie und Sozialpsychologie, 34.*

EPSTEIN, A. L. (1978). *Ethos and identity. Three studies in ethnicity.* London: Tavistock.

EREZ, E. (1984). Self-defined 'desert'and citizen's assessment of the police. *Journal of Criminal Law and Criminology, 75,* 1276-1299.

ERICSSON, K. A., & SIMON, H. A. (1984). *Verbal reports as data.* Cambridge, MA: MIT Press.

ESSED, P. J. M. (1984). *Alledaags racisme* [Every-day racism]. Amsterdam: Sara. (English translation in preparation).

ESSED, P. J. M. (1988). Understanding verbal accounts of racism: Politics and heuristics of reality constructions. In T. A. van Dijk & R. Wodak (Eds.), *Discourse, racism and social psychology.* Special issue of Text, nr.8, 5-40.

FAEGIN, J. R. (1987). Changing black Americans to fit a racist system? *Journal of Social Issues, 43,* 85-89.

FARLEY, R. (1977). Trends in racial inequalities: Have the gains of the 1960s disappeared in the 1970s? *American Sociological Review, 42,* 189-208.

FARLEY, R. (1984). *Blacks and whites: Narrowing the gap?* Cambridge, MA: Harvard University Press.

FARLEY, R. (1985). Three steps forward and two back? Recent changes in the social and economic status of blacks. *Ethnic and Racial Studies, 8,* 4-28.

FELDMAN-SUMMERS, S., & KIESLER, S. B. (1974). Those who are number two try harder: The effect of sex on attributions of causality. *Journal of Personality and Social Psychology, 30,* 846-855.

FERNANDEZ, J. P. (1982). *Racism and sexism in corporate life: changing values in American business.* Lexington, MA: Heath.

FESTINGER, L. (1954). A theory of social comparison processes. *Human Relations, 7,* 117-140.

FESTINGER, L. (1957). *A theory of cognitive dissonance.* Stanford, CA: Stanford University Press.

FIEDLER, F. E. (1967). *A theory of leadership effectiveness.* New York: McGraw-Hill.

FIEDLER, F. E., MITCHELL, T., & TRIANDIS, H. C. (1971). The Culture Assimilator: An approach to cross-cultural training. *Journal of Applied Psychology, 55,* 95-102.

FIENBERG, S. E. (1977). *The analysis of cross-classified categorical data.* London (England) & Cambridge MA: The MIT-Press.

FISKE, S. T. (1982). Schema-triggered affect: Applications to social perception. In M. S. Clark & S. T. Fiske (Eds.), *Affect and cognition: The 17th annual carnegie symposium on cognition* (pp. 55-78). Hillsdale, NJ: Erlbaum.

FISKE, S. T., & TAYLOR, S. E. (1984). *Social cognition.* Reading, MA: Addison-Wesley.

FOWLER, R., HODGE, B., KRESS, G., & TREW, T. (1979). *Language and control.* London: Routledge & Kegan Paul.

FREEMAN, R. (1978). Black economic progress since 1964. *Public Interest, 52,* 52-69.

FRENCH, V. V. (1947). The structure of sentiments I: A restatement of the theory of sentiments. *Journal of Personality, 15,* 247-282.

FRIJDA, N. H. (1986). *The emotions.* Cambridge University Press, New York.

GAERTNER, S. L. (1976). Nonreactive measures in racial attitude: A focus on 'liberals'. In P. A. Katz (Ed.), *Toward the elimination of racism.* New York: Pergamon.

GAERTNER, S. L., & DOVIDIO, J. F. (1977). The subtlety of white racism, arousal, and helping behavior. *Journal of Personality and Social Psychology, 35,* 691-707.

GAERTNER, S. L., & DOVIDIO, J. F. (1986). Prejudice, discrimination, and racism: problems, progress, and promise. In J. F. Dovidio and S. L. Gaertner (Eds). *Prejudice, discrimination, and racism.* New York: Academic Press.

GAERTNER, S. L., & DOVIDIO, J. F.(1986). The aversive form of racism. In J. F. Dovidio and S. L. Gaertner (Eds). *Prejudice, discrimination, and racism.* New York: Academic Press.

GARCIA, L. T., ERSKINE, N., HAWN, K., & CASMAY, S. R. (1981). The effect of affirmative action on attributions about minority group members. *Journal of Personality, 49,* 427-437.

GERARD, H. B., & HOYT, M. F. (1974). Distinctiveness of social categorization and attitude towards ingroup members. *Journal of Personality and Social Psychology, 29,* 836-842.

GLASER, B. G., & STRAUSS, A. L. (1967). *The discovery of grounded theory.* Chicago: Aldine.

GLAZER, N. AND MOYNIHAN, D. (1963). *Beyond the melting pot.* Cambridge, MA: MIT Press.

GREENBERG, J., & ROSENFIELD, D. (1979). White's ethnocentrism and their attributions for the behaviour of blacks: A motivational bias. *Journal of Personality, 47,* 643-657.

GURIN, P., GURIN, G., LAO, R., & BEATTIE, H. (1969). Internal-external control in the motivational dynamics of negro youth. *Journal of Social Issues, 25,* 29-53.

GURR, T. R. (1968). Psychological factors in civil violence. *World Politics, 20,* 245-278.

GURR, T. R. (1970). *Why men rebel*. Princeton, NJ: Princeton University Press.

GURWITZ, S. B., & DODGE, K. A. (1977). Effects of confirmations and disconfirmations on stereotype-based attributions. *Journal of Personality and Social Psychology, 35*, 495-500.

GUTEK, B., LARWOOD, L., & STROMBERG, A. (1985). Women at work. In C. Cooper & I. Robertson (Eds.), *Review of Industrial/Organizational Psychology* (Vol. 1). New York: Wiley.

HACKMAN, R., & OLDHAM, G. R. (1976). Motivation through the design of work. *Organizational Behavior and Human Performance, 16*, 250-279.

HAMILTON, D. L. (1979). A cognitive-attributional analysis of stereotyping. In L. Berkowitz (Ed.), *Advances in experimental social psychology* (Vol. 12). New York: Academic.

HAMILTON, D. L. (Ed.) (1981). *Cognitive processes in stereotyping and intergroup behavior*. Hillsdale, MA: Erlbaum.

HAMILTON, D. L., & TROLIER, T. K. (1986). Stereotypes and stereotyping: an overview of the cognitive approach. In J. F. Dovidio & S. L. Gaertner (Eds.), *Prejudice, discrimination and racism*. Orlando, Fl.: Academic Press.

HAMNER, W. C., KIM, J. S., BAIRD, L., & BIGONESS, W. J. (1974) Race and sex as determinants of ratings by potential employers in a simulated work sampling task. *Journal of Applied Psychology, 59*, 705-711.

HARDING, J., PROSHANSKY, H., KUTNER, B., & CHEIN, I. (1954). Prejudice and ethnic relations. In G. Lindzey & E. Aronson (Eds.), *The handbook of social psychology* (2nd ed.). Vol. V. Reading, MA: Addison-Wesley

HARRÉ, R., & SECORD, P. F. (1972). *The explanation of social behavior*. Oxford: Blackwell.

HARVEY O. J. (1956). An experimental investigation of negative and positive relations between small groups through judgmental indices. *Sociometry, 19*, 201-209.

HAUSER, R. M., & FEATHERMAN, D. L. (1974). White-nonwhite differentials in occupational mobility among men in the United States, 1962-1972. *Demography, 11*, 247-265.

HEILMAN, M. E. (1979). High school students'occupational interest as a function of projected sex ratios in male-dominated occupations. *Journal of Applied Psychology, 64*, 275-279.

HEILMAN, M. E., & HERLIHY, J. M. (1984). Affirmative action, negative reaction? Some moderating conditions. *Organizational Behavior and Human Performance, 33*, 204-213.

HENDRICKS, M., & BOOTZIN, R. (1976). Race and sex as stimuli for negative affect and physical avoidance. *Journal of Social Psychology, 98*, 111-120.

HEPBURN, J. R. (1978). Race and the decision to arrest: an analysis of warrants issued. *Journal of Research in Crime and Delinquency, 15*, 54-73.

HEWSTONE, M., BENN, W., & WILSON, A. (1988). Bias in the use of base rates: Racial prejudice in decision making. *European Journal of Social Psychology., 18*, 161-176.

HEWSTONE, M., BOND, M. H., & WAN, K. -C. (1983). Social facts and social attributions: The explanation of intergroup differences in Hong Kong. *Social Cognition, 2*, 140-155.

HEWSTONE, M., & BROWN, R. J. (1986). Contact is not enough: An intergroup perspective on the "contact hypothesis". In M. Hewstone & R. J. Brown (Eds.), *Contact and Conflict in intergroup encounters*. Oxford: Blackwell.

HEWSTONE, M., & JASPARS, J. (1982a). Intergroup relations and attribution processes. In H. Tajfel (Ed.), *Social identity and intergroup relations*. Cambridge/Paris: Cambridge University Press, Maison des Sciences de l'Homme.

HEWSTONE, M., & JASPARS, J. (1982b). Explanations for racial discrimination: The effect of group discussion on intergroup attributions. *European Journal of Social Psychology, 12*, 1-16.

HEWSTONE, M., & JASPARS, J. (1984). Social dimensions of attribution. In H. Tajfel (Ed.), *The social dimension: European developments in social psychology*. Cambridge/Paris: Cambridge University Press/Maison des Sciences de l'Homme.

HEWSTONE, M., JASPARS, J., & LALLJEE, M. (1982). Social representations, social attribution and social identity: The intergroup images of "public" and "comprehensive" schoolboys. *European Journal of Social Psychology, 12*, 241-269.

HEWSTONE, M., & WARD, C. (1985). Ethnocentrism and causal attribution in Southeast Asia. *Journal of Personality and Social Psychology, 48*, 614-623.

HILL, G. D., HARRIS, A. R., & MILLER, J. L. (1985). The etiology of bias: social heuristics and rational decision making in deviance processing. *Journal of Research in Crime and Delinquency, 22*, 135-162.

HIRSCHHORN, E. (1976). Federal legal remedies for racial discrimination. In P. A. Katz (Ed.), *Towards the elimination of racism* (pp. 377-440). New York: Pergamon.

HOLZKAMP, K. (1973). *Sinnliche Erkenntnis. Historischer Ursprung und gesellschafliche Funktion der Wahrnehmung* [Sensory insight. Historical origin and societal function of the senses]. Frankfurt am Main: Fischer Taschenbuch.

HOUT, M. (1984). Occupational mobility of black men: 1962 to 1973. *American Sociological Review, 49*, 308-322.

HUGHES, R. (1986). *The fatal shore*. New York: Knopf.

HYMAN, H. H. (1975). *Interviewing in social research*. Chicago: University of Chicago Press.

HYMAN, H. H., & SINGER, E. (Eds.). (1968). *Readings in reference group theory and research*. New York: The Free Press.

ICKES, W. (1984). Compositions in black and white: Determinants of interaction in interracial dyads. *Journal of Personality and Social Psychology, 47*, 330-341.

JAHODA, G. (1961). *White man*. London: Oxford University Press for the Institute of Race Relations.

JAULIN, R. (1973). *Gens du soi, gens de l'autre* [Own people, foreign people]. Paris: Union générale d'Edition.

JOHNSON, D. W., & JOHNSON, R. T. (1975). *Learning together and alone: Cooperation, competition and individualization*. Englewood Cliffs, NJ: Prentice-Hall.

JOHNSON, D. W., & JOHNSON, R. T. (1985). Relationships between black and white students in intergroup cooperation and competition. *The Journal of Social Psychology, 125*, 421-428.

JOHNSON, D. W., JOHNSON, R., & MARUYAMA, G. (1984). Goal interdependence and interpersonal attraction. In N. Miller and M. B. Brewer (Eds.), *Groups in contact: The psychology of desegragation*. New York: Academic Press.

JOHNSON, D. W., JOHNSON, R. AND SMITH, K. A. (1986). Academic conflict among students: controversy and learning. In R. S. Feldman (Ed.), *The social psychology of education*. Cambridge: Cambridge University Press.

JOHNSON, J. T., JEMMOTT, J. B., & PETTIGREW, T. F. (1984). Causal attributions and dispositional inference: Evidence of inconsistent judgments. *Journal of Experimental Social Psychology, 20,* 567-585.

JONES, E. E., & DAVIS, K. E. (1965). From acts to dispositions: The attribution process in person perception. In L. Berkowitz (Ed.), *Advances in experimental social psychology* (Vol. 2) New York: Academic Press.

JONES, J., & OLMEDO, E. (1986). *Bias as a cause of minority turnover.* Unpublished paper. Office of Ethnic Minority Affairs, American Psychological Association, Washington, DC.

JUNGER-TAS, J., & VAN DER ZEE-NEFKENS, A. A. (1977). *Een observatieonderzoek naar politiesurveillance* [An obervational study on police patrolling]. Den Haag: W.O.D.C.

KAGAN, S. (1980). Cooperation-competition, culture, and structural bias in classrooms. In S. Sharan, P. Hare, C. Webb, and R. Hertz-Lazarowitz (Eds.), *Cooperation in education.* Provo, UT: Brigham Young University Press.

KAGAN, S. (1985). Dimensions of cooperative classroom structures. In R. Slavin, S. Sharan, S. Kagan, R. Hertz Lazarowitz, C. Webb, and R. Schmuck, *Learning to cooperate, cooperating to learn.* New York: Plenum Press.

KANTER, R. M. (1977). Some effects of proportions on group life: Skewed sex ratios and responses to token women. *American Journal of Sociology, 82,* 965-991.

KATZ, I. (1979). Some thoughts about the stigma notion. *Personality and Social Psychology Bulletin, 5,* 447-460.

KATZ, P. A. (1976). *Towards the elimination of racism.* New York: Pergamon.

KERR, S. (1975). On the folly of rewarding A, while hoping for B. *Academy of Management Journal, 18,* 769-783.

KIDD, J. W. (1953). Personality traits as barriers to acceptability in a college men's residence hall. *Journal of Social Psychology, 38,* 127-130.

KLECK, R. (1968). Physical stigma and nonverbal cues emitted in face-to-face in interactions. *Human Relations, 21,* 19-28.

KLINEBERG, O., & ZAVALLONI, M. (1969). *Nationalism and tribalism among African students.* The Hague & Paris: Mouton.

KLUEGEL, J. R., & SMITH, E. R. (1986). *Beliefs about inequality: Americans' views or what is and what ought to be.* New York: Aldine de Gruyter.

KONECNI, V. J. (1979). The role of aversive events in the development of intergroup conflict. In W. G. Austin & S. Worchel (Eds.), *The social psychology of intergroup relations.* Monterey, CA: Brooks/Cole.

KONECNI, V. J., & KOVEL, J. (1970). *White racism: A psychological history.* New York: Pantheon.

KRAM, K. E. (1983). Phases of the mentor relationship. *Academy of Management Journal, 26,* 608-625.

KULIK, J. A. (1983). Confirmatory attribution and the perpetuation of social beliefs. *Journal of Personality and Social Psychology, 44,* 1171-1181.

LABOV, W. (1972). The transformation of experience in narrative syntax. In W. Labov, *Language in the inner city* (pp. 354-396). Philadelphia, PA: University of Pennsylvania Press.

LALONDE, R. N., MOGHADDAM, F. M., & TAYLOR, D. M. (1987). The process of group differentiation in a dynamic intergroup setting. *Journal of Social Psychology, 127,* 273-287.

LAMBERT, W. E., HODGSON, R. C., GARDNER, R. C., & FILLENBAUM, S. (1960). Evaluational reactions to spoken languages. *Journal of Abnormal and Social Psychology, 60*, 44-51.

LANDAU, S. F., & NATHAN, G. (1983). Selecting delinquents for cautioning in the London Metropolitan Area. *British Journal of Criminology, 23*, 128-149.

LANDIS, D., DAY, H. R., MCGREW, P. L., THOMAS, J. A., & MILLER, A. B. (1976). Can a Black 'Culture Assimilator' increase racial understanding? *Journal of Social Issues, 32*, 169-183.

LANDIS, D., BRISLIN, R. W., & HULGUS, J. F. (1985). Attributional training versus contact in acculturative learning: A laboratory study. *Journal of Applied Social Psychology, 15*, 466-482.

LANGER, E. J., & ABELSON, R. P. (1974). A patient by any other name: Clinical group difference in labeling bias. *Journal of Consulting and Clinical Psychology, 42*, 4-9.

LARWOOD, L. (1982). The importance of being right when you think you are -An examination of self-serving bias in equal employment opportunity. In B. Gulik (Ed.), *Sex role stereotyping and affirmative action policy*. Los Angeles: Institute of Industrial Relations, U.C.L.A.

LAZARUS, R. S. (1966). *Psychological stress and the coping process*. New York: McGraw-Hill.

LAZARUS, R. S. (1982). Thoughts on the relations between emotion and cognition. *American Psychologist, 37*, 1019-1024.

LAZARUS, R. S. (1984). On the primacy of cognition. *American Psychologist, 39*, 124-12.

LEMAINE, G. (1974). Social differentiation and social originality. *European Journal of Social Psychology, 4*, 17-52.

LEMYRE, L., & SMITH, P. M. (1985). Intergroup discrimination and self-esteem in the minimal group paradigm. *Journal of Personality and Social Psychology, 49*, 660-670.

LEVELT, W. J. M. (1983). Monitoring and self-repair in speech. *Cognition 14*, 41-104.

LEVENTHAL, G. S. (1970). Influence of brother and sisters on sex-role behaviors. *Journal of Personality and Social Psychology, 16*, 452-465.

LÉVI-STRAUSS, C. (1962). *La pensée sauvage* [The unrestricted mind]. Paris: Plon.

LEWIN, K. (1948). *Resolving social conflicts*. New York: Harper & Bros.

LINVILLE, P. W., & JONES, E. E. (1980). Polarized appraisals of out-group members. *Journal of Personality and Social Psychology, 38*, 689-703.

LIPPMAN, W. (1922). *Public opinion*. New York: Harcourt, Brace.

LOCKSLEY, A., HEPBURN, C., & ORTIZ, V. (1982). Social stereotypes and judgments of individuals: in instance of base-rate fallacy. *Journal of Experimental Social Psychology, 18*, 23-42.

LORENZI-CIOLDI, F. (1988). *Individus dominants et groupes dominés, images masculines et féminines* [Dominant individuals and dominated groups, male and female images]. Grenoble: Presses Universitaires.

LORWIN, V. R. (1972). Linguistic pluralism and political tension in modern Belgium. In J. A. Fishman (Ed.), *Advances in the sociology of language II* (pp. 386-412). Paris, Mouton.

LUNDMAN, R. J., SYKES, R. E., & CLARCK, J. P. (1978). Police control of juveniles: a replication. *Journal of Research in Crime and Delinquency, 15.*

MALPASS, R. S., & KRAVITZ, J. (1969). Recognition for faces of own and other race. *Journal of Personality and Social Psychology, 13,* 330-334.

MANN, J. W. (1961). Group relations in hierarchies. *The Journal of Social Psychology, 54,* 283-314.

MARIN, G., & TRIANDIS, H. C. (1985). Allocentrism as an important characteristic of the behavior of Latin Americans and Hispanics. In R. Diaz-Guerrero (Ed.), *Cross-cultural and national studies in social psychology,* vol. 2, (pp. 85-104). Amsterdam: Elsevier Science.

MARQUES, J. M. (1986). *Toward a definition of social processing of information: An application to stereotyping.* Catholic university of Louvain-La-Neuve, Unpublished doctoral dissertation.

MARTIN, J. (1981). Relative deprivation: A theory of distributive injustice for an era of shrinking resources. *Research in Organizational Behavior, 3,* 53-107.

MARTIN, J., PRICE, R., BIES, R., & POWERS, M. (In press). Now that I can have it, I'm not so sure I want it: The effects of opportunity on aspirations and discontent. In B. Gutek & L. Larwood (Eds.), *Women's career development.* Beverley Hills, CA: Sage.

McCONAHAY, J. B. (1982) Is it the buses or the blacks? Self-interest versus symbolic racism as predictors of opposition to busing in Louisville. *Journal of Politics, 44,* 692-720.

MEHRABIAN, A. (1968). Relationship of attitudes to seated posture, orientation and distance. *Journal of Personality and Social Psychology, 10,* 26-30.

MERTON, R. K. (1957). *Social theory and social structure.* Glencoe, IL: Free Press.

MERTON, R. K., & KITT, A. S. (1950). Contributions to the theory of reference group behavior. In R. K. Merton & P. F. Lazarsfeld (Eds.), *Continuities in social research, studies in the scope and method of "The American soldier".* Glencoe, IL.: The Free Press.

MILLER, A. G. (Ed.) (1982). *In the eye of the beholder: Contemporary issues in stereotyping.* New York: Praeger.

MILLER, D. T., & ROSS, M. (1975). Self-serving biases in the attribution of causality: Fact or fiction? *Psychological Bulletin, 82,* 213-225.

MILLER, N., & BREWER, M. B. (Eds.) (1984). *Groups in contact: The psychology of desegregation.* Orlando: Academic Press.

MILNER, D. (1983). *Children and race. Ten years on.* London: Wardlock Educational.

MITCHELL, T., DOSSETT, D. L., FIEDLER, F. E., & TRIANDIS, H. C. (1972). Culture training: Validation evidence for the culture assimilator. *International Journal of Psychology, 7,* 97-104.

MOGHADDAM, F. M., & STRINGER, P. (1986). Trivial and important criteria for social categorization in the minimal group paradigm. *The Journal of Social Psychology, 126,* 345-354.

MOSCOVICI, S., & PAICHELER, G. (1978). Social comparison and social recognition: Two complementary processes of identification. In H. Tajfel (Ed.), *Differentiation between groups: Studies in the social psychology of intergroup relations.* London: Academic Press.

MUMMENDEY, A., & SCHREIBER, H. S. (1984). 'Different' just means 'better': Some obvious and some hidden pathways to ingroup favouritism. *British Journal of Social Psychology, 23,* 363-368.

MURRAY, H. A. (1933). The effect of fear upon estimates of the maliciousness of other personalities. *Journal of Social Psychology, 4,* 310-329.

MYERS, D. G., & BISHOP, G. D. (1970). Discussion effects on racial attitudes. *Science, 169,* 778-789.

NAVON, D. (1977). Forest before trees: the precedence of global features in visual perception. *Cognitive Psychology, 9,* 353-383.

NEWCOMB, T. M. (1961). *The acquaintance process.* New York: Holt, Rinehart & Winston.

NG, S. H. (1986). Equity, intergroup bias and interpersonal bias in reward allocation. *European Journal of Social Psychology, 16,* 239-255.

NICKERSON, S., MAYO, C., & SMITH, A. (1986). Racism in the courtroom. In J. F. Dovidio & S. L. Gaertner (Eds.), *Prejudice, discrimination and racism.* Academic Press, 1986.

NISBETT, R. E., & WILSON, T. D. (1977). Telling more than we can know: Verbal reports on mental processes. *Psychological Review, 84,* 231-259.

NORTHCRAFT, G. B. (1982). *Affirmative action: The impact of legislated equality.* Unpublished doctoral dissertation, Stanford University.

NORTHCRAFT, G. B., & MARTIN, J. (1982). Double jeopardy: Resistance to affirmative action from potential beneficiaries. In B. A. Gutek (Ed.), *Sex role stereotyping and affirmative action policy.* Los Angeles: Institute of Industrial Relations, U.C.L.A.

NOVACO, R. W. (1979). The cognitive regulation of anger and stress. In P. C. Kendall & S. D. Hollon (Eds.), *Cognitive-behavioural interventions: theory, research, and procedures.* New York: Academic Press.

OSKAMP, S. *Applied social psychology.* Englewood Cliffs: Prentice-Hall, 1984.

PARK, B., & ROTHBART, M. (1982). Perception of out-group homogeneity and levels of social categorization: memory for the subordinate attributes of in-group and out-group members. *Journal of Personality and Social Psychology, 42,* 1051-1068.

PARK, R. E. (1928/1950). The basis of race prejudice. The Annals, CXXXX, 11-20. (Reprinted in R. E. Park (1950), *Race and culture.* Glencoe, IL: The Free Press. )

PATCHEN, M. (1982). *Black-white contact in schools: its social and academic effects.* Lafayette, IN: Purdue University Press.

PEABODY, D. (1985). *National characteristics.* Cambridge: Cambridge University Press.

PETTIGREW, T. F. (1958). Personality and sociocultural factors in intergroup attitudes: A cross-cultural comparison. *Journal of Conflict Resolution, 2,* 29-42.

PETTIGREW, T. F. (1959). Regional differences in anti-Negro prejudice. *Journal of Abnormal and Social Psychology, 59,* 28-36.

PETTIGREW, T. F. (1961). Social psychology and desegregation research. *American Psychologist, 16,* 105-112.

PETTIGREW, T. F. (1967). Social evaluation theory: convergences and applications. *Nebraska Symposium on Motivation* (pp. 241-318). Nebraska: University of Nebraska Press.

PETTIGREW, T. F. (1971). *Racially separate or together?* New York: McGraw-Hill.

PETTIGREW, T. F. (1975). The racial integration of the schools. In T. F. Pettigrew (Ed.), *Racial discrimination in the United States.* New York: Harper & Row.

PETTIGREW, T. F. (1979). The ultimate attribution error: Extending Allport's cognitive analysis of prejudice. *Personality and Social Psychology Bulletin, 5,* 461-476.

PETTIGREW, T. F. (1979). Racial change and social policy. *Annals of the American Academy of Political and Social Science, 441,* 114-131.

PETTIGREW, T. F. (1986). The intergroup contact hypothesis reconsidered. In M. Hewstone & R. Brown (Eds.), *Contact, conflict and intergroup relations.* Oxford, England: Blackwell.

PETTIGREW, T. F. (1989). *Modern racism: American black-white relations since the 1960s.* Cambridge, MA: Harvard University Press.

PETTIGREW, T. F., JEMMOTT, J. B., & JOHNSON, J. T. (1986). *Race and the questioner effect: Testing the ultimate attribution error.* Unpublished paper, Dept. of Psychology, University of California, Santa Cruz.

PETTIGREW, T. F., & MARTIN, J. (1987). Shaping the organizational context for black american inclusion. *Journal of Social Issues, 43,* 41-78.

PLECK, J. H. (1977). The work-family role system. *Social Problems, 24,* 417-427.

PLUMMER, K. (1983). *Documents of life. An introduction to the problems and literature of a humanistic method.* London: Allen & Unwin.

PORTER, L. W., & LAWLER, E. E. (1968). *Managerial attitudes and performance.* Homewood, IL: Irwin.

POTTER, J., & WETHERELL, M. (1987). *Discourse and social psychology.* London: Sage.

POTTER, J., & WETHERELL, M. (1988). Accomplishing attitudes: Fact and evaluation in racist discourse. In T. A. van Dijk & R. Wodak (Eds.), *Discourse, racism and social psychology.* Special issue of Text 8, nr. 1, 51-68.

PYSZCZYNSKI, T. A., & GREENBERG, J. (1981). Role of disconfirmed expectancies in the instigation of attributional processing. *Journal of Personality and Social Psychology, 45,* 323-334.

QUATTRONE, G. A. (1986). On the perception of a group's variability. In S. Worchel and W. G. Austin (Eds.), *Psychology of Intergroup Relations.* Chicago: Nelson-Hall.

RABBIE J. M. (1982). The effects of intergroup competition and cooperation on intragroup and intergroup relationships. In V. J. Derlega, J. Grzelak (Ed.), *Cooperation and helping behavior* (pp 123-149). London: Academic Press.

RABBIE J. M., & BEKKERS, F. (1976). Bedreigd leiderschap en intergroep competitie [Threatened leadership and intergroup competition]. *Nederlands Tijdschrift voor de Psychologie, 31,* 269-283.

RANDOLPH, G., LANDIS, D., & TZENG, O. C. S. (1977). The effects of time and practice upon culture assimilator training. *International Journal of Intercultural Relations, 1,* 105-119.

RASINSKI, K. A., CROCKER, J., & HASTIE, R. (1985). Another look at sex stereotypes and social judgments: an analysis of the social perceiver's use of subjective probabilities. *Journal of Personality and Social Psychology, 49,* 317-326.

REX, J. (1981). Urban segregation and inner city policy in Great Britain. In C. Peach, V. Robinson en S. Smith (Eds.), *Ethnic segregation in cities* (pp. 25-42). London: Croom Helm.

RIJSMAN, J. (1980). Sociale motivatie [Social motivation]. In R. van der Vlist & J. Jaspars (Eds.), *Sociale Psychologie in Nederland* [Social Psychology in the Netherlands], Deel 1. Deventer: Van Loghum Slaterus.

ROGERS, R. W., & PRENTICE-DUNN, S. (1981). Deindividuation and anger-mediated interracial aggression: Unmasking regressive racism. *Journal of Personality and Social Psychology, 41,* 63-73.

ROKEACH, M. (1960). *The open and closed mind.* New York: Basic Books.

ROKEACH, M. AND MEZEI, L. (1966). Race and shared belief as factors in social choice, *Science, 151,* 167-172.

ROSCH, E. (1978). Principles of categorization. In E.Rosch & B. B. Lloyd (Eds.), *Cognition and Categorization.* Hillsdale, NJ: Erlbaum.

ROSCH, E., MERVIS, C. B., GRAY, W. D., JOHNSON, D. M., & BOYES-BRAEM, P. (1976). Basic objects in natural categories. *Cognitive Psychology, 8,* 382-439.

ROSE, T. L. (1981). Cognitive and dyadic processes in intergroup contact. In D. L. Hamilton (Ed.), *Cognitive processes in stereotyping and intergroup behavior.* Hillsdale, NJ: Erlbaum.

ROSENTHAL, R., & ROSNOW, R. (Eds.) (1969). *Artifact in behavioral research.* New York: Academic Press.

ROSS, E. A. (1920). *The principles of sociology.* New-York: The Century Co.

ROSS, G. F. (1975). *An experimental investigation of open and closed groups.* University of Bristol: unpublished manuscript.

ROSS, L. D. (1977). The intuitive psychologist and his shortcomings: Distortions in the attribution process. In L. Berkowitz (Ed.), *Advances in Experimental Social Psychology, 10.* New York: Academic Press.

ROSS, L. D., AMABILE, T. M., & STEINMETZ, J. L. (1977). Social roles, social control, and biases in social-perception processes. *Journal of Personality and Social Psychology, 35,* 485-494.

ROTHBART, M. (1981). Memory processes and social beliefs. In D. L. Hamilton (Ed.), *Cognitive processes in stereotyping and intergroup behavior.* Hillsdale, NJ: Erlbaum.

ROTHBART, M., & JOHN, O. P. (1985). Social categorization and behavioral episodes: A cognitive analysis of the effects of intergroup contact. *Journal of Social Issues, 41,* 81-104.

ROTTER, J. B. (1966). Generalized expectancies for internal versus external control of reinforcement. *Psychological Monographs, 80* (1, Whole No. 609).

RUNCIMAN, W.G. (1966). Relative deprivation and social justice. London: Routledge and Kegan Paul.

SACHDEV, I., & BOURHIS, R. Y. (1984). Minimal majorities and minorities. *European Journal of Social Psychology, 14,* 35-52.

SACHDEV, I., & BOURHIS, R. Y. (1985). Social categorization and power differentials in group relations. *European Journal of Social Psychology, 15,* 415-434.

SACHDEV, I., & BOURHIS, R. Y. (1987). Status differentials and intergroup behaviour. *European Journal of Social Psychology, 17,* 277-293.

SAGAR, H. A., & SCHOFIELD, J. W. (1980). Racial and behavioral cues in black and white children's perceptions of ambiguously aggressive acts. *Journal of Personality and Social Psychology, 39,* 590-598.

SARTRE, J. -P. (1946). *Réflexions sur la question Juive* [Anti-Semite and Jew]. Paul Morihien, Paris.

SCHLENKER, B. R. (1975). Self-presentation: Managing the impression of consistency when reality interferes with self-enhancement. *Journal of Personality and Social Psychology, 32,* 1030-1037.

SCHOFIELD, J. W. (1982). *Black and white in school: Trust, tension or tolerance?* New York: Praeger.

SCHOFIELD, J. W. (1986). Black-white contact in desegregated schools. In M. Hewstone and R. Brown (Eds.), *Contact & conflict in intergroup encounters*. New York: Basil Blackwell.

SCHOFIELD, J. W., SCHÖNBACH, P., GOLLWITZER, P., STIEPEL, G., & WAGNER, U. (1981). *Education and intergroup attitude*. New York: Academic Press.

SCHUMAN, H., STEEH, C., & BOBO, L. (1985). *Racial attitudes in America: Trends and interpretations*. Cambridge, MA: Harvard University Press.

SCHWARTZ, H., & JACOBS, J. (1979). *Qualitative sociology*. New York: Free Press.

SCHWARZWALD, J., & YINON, Y. (1977). Symmetrical and asymmetrical interethnic perception in Israel. *International Journal of Intercultural Relations, 1*, 40-47.

SHARAN, S. (1980). Cooperative learning in small groups: Recent methods and effects on achievement, attitudes, and ethnic relations. *Review of Educational Research, 50*, 241-271.

SHARAN, S., & HERTZ-LAZAROWITZ, R. (1980). A group-investigation method of cooperative learning in the classroom. In S. Sharan, P. Hare, C. Webb, and R. Hertz-Lazarowitz (Eds.), *Cooperation in education*. Provo, UT: Brigham Young University Press.

SHERIF, M. (1966). *In common predicament*. Boston: Houghton, Mifflin.

SHERIF, M. (1967). *Group conflict and co-operation*. London: Routledge and Kegan Paul.

SHERIF, M., HARVEY, O. J., WHITE, B. J., HOOD, W. R., & SHERIF, C. W. (1961). *Intergroup conflict and co-operation. The Robbers Cave experiment*. Norman, OK: University of Oklahoma Book Exchange.

SHERIF, M., & SHERIF, C. W. (1969). *Social psychology*. New York: Harper & Row.

SHERWOOD, G. G. (1981). Self-serving biases in person perception: A reexamination of projection as a mechanism of defense. *Psychological Bulletin, 90*, 445-459.

SHOEMAKER, D. J., SOUTH, D. R., & LOWE, J. (1972). Facial stereotypes of deviants and judgments of guilt or innocence. *Social Forces, 51*, 427-433.

SIMMEL, G. (1971). *On individuality and social forms*. Chicago: University of Chicago Press.

SIMPSON, G. E., & YINGER, J. M. (1986). *Racial and cultural minorities. An analysis of prejudice and discrimination*. New York: Plenum Press.

Slavin, R. E. (1980). Cooperative learning. *Review of Educational Research, 50*, 315-342.

SLAVIN, R. E. (1983). *Cooperative learning*. New York: Longman.

SLAVIN, R. E. (1985). Cooperative learning: Applying contact theory in desegregated schools. *Journal of Social Issues, 41*, No 3, 45-62.

SIMMONS, R. (1978). Blacks and high self-esteem: A puzzle. *Social Psychology, 41*, 54-57.

SMIT, V., & NOORT, K. (1987). *Crooswijk van "bijzonder" naar "gewoon"? Onderzoek naar de effecten van het stadsvernieuwingsproces op de bevolkingsstructuur van de Rotterdamse Crooswijk* [Research on the effects of an urban renewal process on the demographic composition of a Rotterdam city quarter]. Eindhoven: T.U. Eindhoven afdeling Bouwkunde.

SMITH, D. A., VISHER, C. A., & DAVIDSON, L. A. (1984). Equity and discretionary justice: the influence of race on police arrest decisions. *Journal of Criminal Law and Criminology, 75*, 234-249.

SMITH, L. (1949). *Killers of the dream*. New York: Norton.

SMITH, T. W., & SHEATSLEY, P. B. (1984, Oct./Nov.). American attitudes toward race relations. *Public Opinion, 7*, 14-15, 50-53.

SNOODGRASS, S. E. (1985). Women's intuition: the effect of subordinate role on interpersonal sensitivity. *Journal of Personality and Social Psychology, 49*, 146-155.

SNYDER, M., & CANTOR, N. (1979). Testing hypotheses about other people: The use of historical knowledge. *Journal of Experimental Social Psychology, 15*, 330-343.

SOLOMAN, R. L. (1964). Punishment. *American Psychologist, 19*, 239-253.

SPRADLEY, J. P. (1979). *The ethnographic interview*. New York: Holt, Rinehart & Winston.

SROUFE, L. A. (1977). Wariness of strangers and the study of child development. *Child Development, 48*, 731-746.

SROUFE, L. A. (1979). Socioemotional development. In J. Osofsky (Ed.), *Handbook of Infant Development*. New York: Wiley.

STAROBINSKI, J. (1964). *L'Invention de la liberté* [The invention of liberty]. Genève: Skira.

STEPHAN, W. G. (1985). Intergroup relations. In G. Lindzey & E. Aronson (Eds.), *The handbook of social psychology* (3rd ed.) Vol. II. New York: Random House.

STEPHAN, W. G. (1987). The contact hypothesis in intergroup relations. In C. Hendrick (Ed.), *Group processes and interpersonal relations* (pp. 13-40).

STEPHAN, W. G., & ROSENFIELD, D. (1982). Racial and ethnic stereotyping. In A. G. Millar (Ed.), *In the eye of the beholder: Contemporary issues in stereotyping*. New York: Praeger.

STEPHAN, W. G., & STEPHAN, C. W. (1984). The role of ignorance in intergroup relations. In N. Miller & M. B. Brewer (Eds.), *Groups in contact: The psychology of desegregation*. Orlando: Academic Press.

STEPHAN, W. G., & STEPHAN, C. W. (1985). Intergroup anxiety. *Journal of Social Issues, 41*, 157-175.

STOUFFER, S. A., SUCHMAN, E. A., DEVINNEY, L. C., STAR, S. A., & WILLIAMS, R. M. (1949). *The American soldier: Adjustment during army life, Vol. 1*. Princeton NJ: Princeton University Press.

SUMNER, W. G. (1906). *Folkways*. Boston: Ginn.

TAJFEL, H. (1959). A note on Lambert's evaluational reactions to spoken languages. *Canadian Journal of Psychology 13*, 86-92.

TAJFEL, H. (1969). Cognitive aspects of prejudice. *Journal of Social Issues, 25*, 79-97.

TAJFEL, H. (1970). Experiments in intergroup discrimination. *Scientific American, 223*, 96-102.

TAJFEL, H. (1972). La catégorisation sociale [Social categorization]. In Moscovici, S. (Ed.), *Introduction à la psychologie sociale* [Introduction to Social Psychology], *Vol. 1*. Paris: Larousse.

TAJFEL, H. (1974). Social identity and intergroup behaviour. *Social Science Information, 13*, 65-93.

TAJFEL, H. (1978). *The social psychology of minorities*. London: Minority Right Group Report, 38.

TAJFEL, H. (1978). *Differentiation between social groups.* London: Academic Press.

TAJFEL, H. (1981). *Human groups and social categories.* Cambridge: Cambridge University Press.

TAJFEL, H. (Ed.) (1982). *Social identity and intergroup relations.* Cambridge: Cambridge University Press.

TAJFEL, H. (1982). Social psychology of intergroup relations. *Annual Review of Psychology, 33,* 1-39.

TAJFEL, H., BILLIG, M., BUNDY, R., & FLAMENT, C. (1971). Social categorization and intergroup behaviour. *European Journal of Social Psychology, 1,* 149-178.

TAJFEL, H., SHEIKH, A. A., & GARDNER, R. C. (1964). Content of stereotypes and the inference of similarity between members of stereotyped groups. *Acta Psychologica, 22,* 191-202.

TAJFEL, H., & TURNER, J. C. (1979). *An integrative theory of intergroup conflict. The social psychology of intergroup relations.* Monterey, CA: Brooks/Cole.

TAJFEL, H., & TURNER, J. C. (1985). The social identity theory of intergroup behaviour. In S. Worchel and W. G. Austin (Eds.), *Psychology of intergroup relations* (pp. 7-24). Chicago: Nelson-Hall.

TAJFEL, H., & WILKES, A. L. (1963). Classification and quantitative judgment. *British Journal of Psychology, 54,* 101-114.

TAYLOR, D. G., SHEATSLEY, P. B., & GREELEY, A. M. (1978, June). Attitudes toward racial integration. *Scientific American, 238,* 42-51.

TAYLOR, D. M., & JAGGI, V. (1974). Ethnocentrism and causal attribution in a south India context. *Journal of Cross-Cultural Psychology, 5,* 162-171.

TAYLOR, D. M., & McKIRNAN, D. J. (1984). A five-stage model of intergroup relations. *British Journal of Social Psychology, 23,* 291-300.

TAYLOR, D. M., & MOGHADDAM, F. M. (1987). *Theories of intergroup relations: International social psychological perspectives.* New York: Praeger.

TAYLOR, S. E., & FISKE, S. T. (1978). Point-of-view and perceptions of causality. *Journal of Personality and Social Psychology, 32,* 439-445.

TAYLOR, S. E., & FISKE, S. T. (1978). Salience, attention and attribution: Top of the head phenomena. In L. Berkowitz (Ed.), *Advances in experimental social psychology* (Vol. 11). New York: Academic Press.

TAYLOR, S. E., & FISKE, S. T., CLOSE, M., ANDERSON, C., RUDERMAN, A. (1977). *Solo status as a psychological variable: The power of being distinctive.* Unpublished paper, Harvard University.

TAYLOR, S. E., FISKE, S. T., ETCOFF, N., & RUDERMAN, A. (1978). The categorical and contextual bases of person memory and stereotyping. *Journal of Personality and Social Psychology, 36,* 778-793.

TAYLOR, S. E., & KOIVUMAKI, J. H. (1976). The perception of self and others: Acquaintanceship, affect, and actor-observer differences. *Journal of Personality and Social Psychology, 33,* 403-408.

TESSER, A., & CAMPBELL, J. (1983). Self-definition and self-evaluation maintenance. In J. Suls and A. Greenwald (Eds.), *Social psychological perspectives on the self* (Vol. 2). Hillsdale, NJ: Erlbaum.

TETLOCK, P. E., & LEVI, A. (1982). Attribution bias: On the inconclusiveness of the cognition-motivation debate. *Journal of Experimental Social Psychology, 18,* 68-88.

THIBAUT, J. W., & KELLEY, H. H. (1959). *The social psychology of groups.* New York: Wiley.

TOUHEY, J. (1974). Effects of additional women professionals on ratings of occupational prestige and desirability. *Journal of Personality and Social Psychology, 29,* 86-89.

TRIANDIS, H. C. (1975). Culture training, cognitive complexity and interpersonal attitudes. In R. W. Brislin, S. Bochner & W. Lonner (Eds.), *Cross-cultural perspectives on learning* (pp. 39-77). Beverley Hills, CA: Sage/Wiley/Halstead.

TURNER, J. C. (1975). Social comparison and social identity: Some prospects for intergroup behaviour. *European Journal of Social Psychology, 5,* 5-34.

TURNER, J. C. (1978). Social comparison, similarity and ingroup favouritism. In H. Tajfel (Ed.), *Differentiation between social groups.* London: Academic Press.

TURNER, J. C. (1981). The experimental social psychology of intergroup behaviour. In J. C. Turner & H. Giles (Eds.), *Intergroup behaviour.* Oxford: Basil Blackwell.

TURNER, J. C. (1987). *Rediscovering the social group: A self-categorization theory.* Oxford: Basil Blackwell.

TURNER, J. C., & BROWN, R. J. (1978). Social status, cognitive alternatives and intergroup relations. In H. Tajfel (Ed.), *Differentiation between social groups; Studies in the social psychology of intergroup relations.* London: Academic Press.

TURNER, J. C., HOGG, M. A., OAKES, P. J., & SMITH, P. M. (1984). Failure and defeat as determinants of group cohesiveness. *British Journal of Social Psychology, 23,* 97-111.

TVERSKY, A., & KAHNEMAN, D. (1974). Judgment under uncertainty: Heuristics and biases. *Science, 185,* 1124-1131.

U.S. DEPARTMENT OF LABOR. (1982). *Employment and earnings.* Washington, DC: U.S. Bureau of Labor Statistics.

VANBESELAERE, N. (1987). The effects of dichotomous and crossed social categorizations upon intergroup discrimination. *European Journal of Social Psychology, 17,* 143-156.

VAN DER HOEVEN, E. (1986). *Allochtone jongeren bij de jeugdpolitie deel 2: een observatieonderzoek* [Non-native juveniles at the juvenile police department, part 2: an observational study]. The Hague, Department of Justice, C.W.O.K., 1986.

VAN DIJK, T. A. (1983). *Minderheden in de media* [Minorities in the media]. Amsterdam: Socialistische Uitgeverij Amsterdam.

VAN DIJK, T. A. (1984). *Prejudice and discourse. An analysis of ethnic prejudice in cognition and conversation.* Amsterdam: Benjamins.

VAN DIJK, T. A. (Ed.) (1985). *Handbook of discourse analysis.* 4 Vols. London: Academic Press.

VAN DIJK, T. A. (1985). Cognitive situation models in discourse production: The expression of ethnic situations in prejudiced discourse. In J. P. Forgas (Ed.), *Language and social situations.* New York: Springer.

VAN DIJK, T. A. (1987). *Schoolvoorbeelden van racisme* [Textbook examples of racism]. Amsterdam: SUA.

VAN DIJK, T. A. (1987). *Communicating racism. Ethnic prejudice in thought and talk.* Newbury Park, CA: Sage.

VAN DIJK, T. A. (1988). *News analysis. Case studies of international and national news in the press.* Hillsdale, NJ: Erlbaum.

VAN DIJK, T. A. (1988). Social cognition, social power and social discourse. In T. A. van Dijk & R. Wodak (Eds.), *Discourse, racism and social psychology*. Special issue of Text 8, nr. 1, 129-157.

VAN DIJK, T. A., & KINTSCH, W. (1983). *Strategies of discourse comprehension*. New York: Academic Press.

VAN KNIPPENBERG, A. (1978). Status differences, comparative relevance and intergroup differentiation. In H. Tajfel (Ed.), *Differentiation between social groups: Studies in the social psychology of intergroup relations* (pp. 171-199). London: Academic Press.

VAN KNIPPENBERG, A. (1984). Intergroup differences in group perceptions. In H. Tajfel (Ed.), *The social dimension: European developments in social psychology* (pp. 560-578). Cambridge: Cambridge University Press.

VAN KNIPPENBERG, A., & WILKE, H. (1979). Perceptions of collégiens and apprentis re-analyzed. *European Journal of Social Psychology, 9,* 427-434.

VAN NIEKERK, M., SUNIER, T., & VERMEULEN, H. (1987). *Onderzoek interetnische relaties* [Research on interethnic relations]. Rijswijk: Ministry of Welfare, Health and Culture.

VAN OUDENHOVEN, J. P., VAN BERKUM, G., & SWEN-KOOPMANS, T. (1987). Effect of cooperation and shared feedback on spelling achievement. *Journal of Educational Psychology, 79,* 92-94.

VAN PRAAG, C. S. (1986). Minderheden vóór en na de nota [Minorities before and after the Bill]. *Migrantenstudies, 2,* nr.4.

VAUGHAN, G. M. (1964). Ethnic awareness in relation to minority group membership. *Journal of Genetic Psychology, 105,* 119-130.

VAUGHAN, G. M. (1978). Social change and intergroup preferences in New Zealand. *European Journal of Social Psychology, 9,* 427-434.

WAGNER, U., LAMPEN, L., & SYLWASSCHY, J. (1986). In-group inferiority, social identity and out-group devaluation in a modified minimal group study. *British Journal of Social Psychology, 25,* 15-23.

WAGNER, U., & SCHONBACH, P. (1984). Links between educational status and prejudice: ethnic attitudes in West Germany. In N. Miller, M. B. Brewer, *Groups in Contact* (pp. 29-52). London: Academic Press.

WALKER, I., & PETTIGREW, T. F. (1984). Relative deprivation theory: An overview and conceptual critique. *British Journal of Social Psychology, 23,* 301-310.

WARD, C., & HEWSTONE, M. (1985). Ethnicity, language and intergroup relations in Malaysia and Singapore: A social psychological analysis. *Journal of Multilingual and Multicultural Development, 6,* 271-296.

WEARY, G. (1979). Self-serving attributional biases: Perceptual or response distortions? *Journal of Personality and Social Psychology, 37,* 1418-1420.

WEBER, R., & CROCKER, J. (1983). Cognitive processes in the revision of stereotype beliefs. *Journal of Personality and Social Psychology, 45,* 961-977.

WEINER, B. (1979). A theory of motivation for some classroom experiences. *Journal of Educational Psychology, 71,* 3-25.

WEINER, B. (1982). The emotional consequences of causal attributions. In M. Clark, & S. T. Fiske (Eds.), *Affect and cognition: The 17th Annual Carnegie Symposium on Cognition*. Hillsdale, NJ: Erlbaum.

WEINER, B. (1983). Some methodological pitfalls in attributional research. *Journal of Educational Psychology, 75,* 530-543.

WEINER, B. (1985). An attributional theory of achievement motivation and emotion. *Psychological Review, 92*, 548-573.

WEINER, B. (1986). *An attributional theory of motivation and emotion.* New York: Springer Verlag.

WEINER, B., FRIEZE, I. H., KUKLA, A., REED, I., REST, S., & ROSENBAUM, R. M. (1972). Perceiving the causes of success and failure. In E. E. Jones, D. E. Kanouse, H. H. Kelley, R. E. Nisbett, S. Valins, & B. Weiner, *Attribution: Perceiving the causes of behavior.* Morristown, NJ: General Learning Press.

WEINREICH, P. (1979). Ethnicity and adolescent identity conflicts. A comparative study. In V. Saifullah Khan (Ed.), *Minority families in Britain.* London: MacMillan.

WEITZ, S. (1972). Attitude, voice and behavior: A repressed affect model of interracial interaction. *Journal of Personality and Social Psychology, 24*, 14-21.

WELDON, D. E., CARLSTON, D. E., RISSMAN, A. K., SLOBODIN, L., & TRIANDIS, H. C. (1975). A laboratory test of effects of culture assimilator training, *Journal of Personality and Social Psychology, 32*, 300-310.

WHEELER, L. (1966). Motivation as a determinant of upward comparison. *Journal of Experimental Social Psychology, Supplement 1*, 27-31.

WHITING, B., & WHITING, J. (1975). *Children of six cultures.* Cambridge, MA: Harvard University Press.

WILDER, D. A. (1984). Intergroup contact: The typical member and the exception to the rule. *Journal of Personality and Social Psychology, 20*, 177-194.

WILDER, D. A., & ALLEN, V. (1974). Effects of social categorization and belief similarity upon intergroup behavior. *Personality and Social Psychology Bulletin, 1*, 281-283.

WILLEMSE, H. M., & MEYBOOM, M. L. (1979). Personal characteristics of 'suspects' and stereotyping by the police: a laboratory experiment. *Abstracts on Police Science, 1979*, 359-367.

WILLIAMS, R. M. (1964). *Strangers next door: Ethnic relations in American communities.* Englewood Cliffs, NJ: Prentice-Hall.

WINKEL, F. W., & KOPPELAAR, L. (1986). Being suspect in cross-cultural interaction: a psychological analysis. *Police Studies, 9*, 125-132.

WISPE, L. G., & FRESHLEY, H. G. (1971). Race, sex, and the sympathetic helping behavior: The broken bag caper. *Journal of Personality and Social Psychology, 17*, 59-65.

WORCHEL S. (1987). *Process and effect of achieving group independence.* Paper presented at the workshop "The psychology of the social group", Office of Naval Research, London (july).

WORCHEL, S., ANDREOLI, V. A., & FOLGER, R. (1977). Intergroup cooperation and intergroup attraction: the effect of previous interaction and outcome of combined effort. *Journal of Experimental Social Psychology, 13*, 131-140.

WORCHEL, S., LEE, J., & ADEWOLE, A. (1975). Effects of supply and demand on object value. *Journal of Personality and Social Psychology, 32*, 906-914.

WORD, C. O., ZANNA, M. P., & COOPER, J. (1974). The nonverbal mediation of self-fulfilling prophecies in interracial interaction. *Journal of Experimental Social Psychology, 10*, 109-120.

WYER JR., R. S., & SRULL, T. K. (Eds.) *Handbook of social cognition.* 3 Vols. Hillsdale, NJ: Erlbaum.

YARKIN, K. L., TOWN, J. P., & WALLSTON, B. S. (1982). Blacks and women must try harder: Stimulus persons' race and sex attributions of causality. *Personality and Social Psychology Bulletin, 8,* 21-24.

ZAJONC, R. B. (1968). Attitudinal effects of mere exposure. *Journal of Personality and social Psychology Monograph Supplement 2, 2,* 2-27.

ZAJONC, R. B. (1980). Feeling and thinking: Preferences need no inferences. *American Psychologist, 35,* 151-175.

ZAJONC, R. B. (1984). On the primacy of affect. *American Psychologist, 39,* 117-123.

ZUCKERMAN, M. (1979). Attribution of success and failure revisited, or: The motivational bias is alive and well in attribution theory. *Journal of Personality, 47,* 245-287.

# Name index

# SUBJECT INDEX

Milton Keynes UK
Ingram Content Group UK Ltd.
UKHW022104141024
449569UK00031B/1775